GW00470533

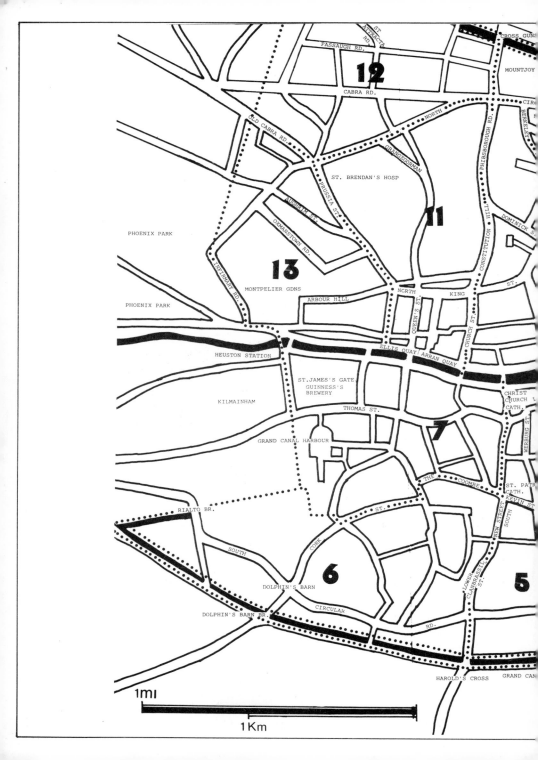

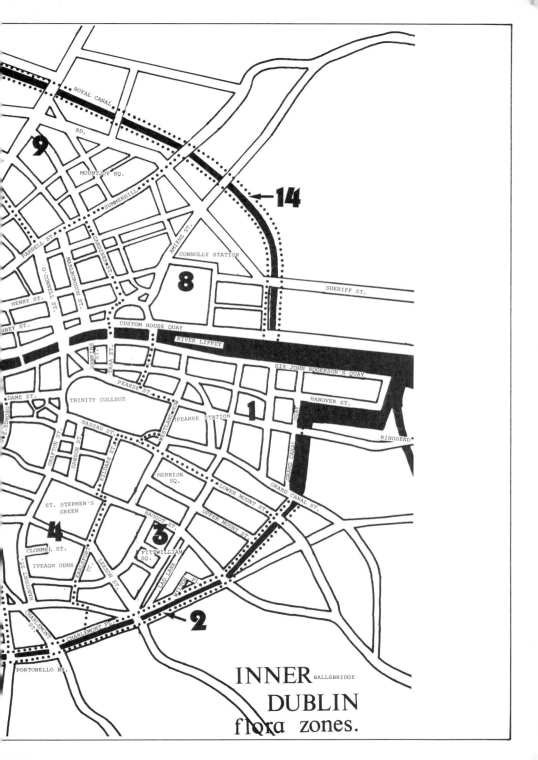

INNER DUBLIN flora zones.

Flora of
Inner Dublin

Flora
of
Inner Dublin

Peter Wyse Jackson
and
Micheline Sheehy Skeffington

Contributions from Declan Doogue, John Akeroyd
and Donal Synnott

and illustrations by Harry McConville and Evelyn Drinan

ROYAL DUBLIN SOCIETY

1984

ISBN   Hardback   0 86027 015 7
        Softback   0 86027 016 5

Science Section,
Royal Dublin Society,
Thomas Prior House,
Ballsbridge, Dublin 4.

Published in association with the Dublin Naturalists' Field Club

Typeset by Print Prep Limited, Dublin

Printed in the Republic of Ireland by Mount Salus Press
Limited, Dublin.

Contents

Acknowledgements

The Royal Dublin Society acknowledges with gratitude assistance in the publication of this Flora from:

An Foras Forbartha
An Taisce — Dublin City Association
Botanical Society of the British Isles
Bryan Guinness Charitable Trust
Department of the Environment — Dublin Inner City Group
Department of Fisheries and Forestry
Dix Transport Limited
Dublin Corporation — Community and Environment Department
Dublin Naturalists' Field Club
Electricity Supply Board
Institute of Biology
Irish Garden Plant Society
Royal Horticultural Society of Ireland
Trinity College Dublin — School of Botany
University College Galway — Department of Botany

The authors would like to thank all the members of the Dublin Naturalists' Field Club who participated with much enthusiasm on field trips and helped in contributing records. A special word of thanks is due to Declan Doogue, John Akeroyd and John Parnell for useful comments on the manuscript and, along with Howard Hudson and Paddy Reilly, for their sustained work in the field, in initial organisation and in collating material.

Thanks are due to Prof. David A. Webb for advice, encouragement and the determination of a number of difficult plant speciments. The authors wish also to acknowledge the special contribution to the project from Maura Scannell, who initially suggested the idea, helped throughout with field work, contributed records, advised on the manuscript and gave continuous encouragement.

We are also grateful to the following:

All zone recorders for their cooperation.

Sylvia Reynolds for help with proof-reading and the index preparation.

Mary Fahy for typing the manuscript.

Prof. Michael E. Mitchell, and Tim Collins for help with the bibliography.

Prof J.F. Killeen for help with Latin translations.

Patrick Wyse Jackson for help with map preparation.

Alan Corsini for preparation and design of the cover.

Dublin Corporation for a grant towards field work expenses.

Preface

I have had the honour to be the President of the Dublin Naturalists' Field Club during the field recording time for the Flora of Inner Dublin (1979-1981) and have pleasure in writing the Preface. Mention is made in the Introduction of the circumstances giving rise to the work and of those responsible for carrying it through. I wish to add my own personal appreciation to those responsible. They have spent long and frequently cold hours in the field accumulating the data here presented, and equally long and tedious hours analysing these data.

This is a rather special Flora as, unlike most of these much loved works of reference, this one is limited to the urban environment of a city and is one of the few of its kind in Ireland and Britain. The problems of the people of the inner city of Dublin have been frequently highlighted in recent years and many changes have taken place in the community life of Dublin city. This is a Flora reflecting change rather than permanency, making it a most valuable record for all present and future botanists. I feel it could act as a guide and a stimulus to workers in other cities.

At the time of publication of Nathaniel Colgan's Flora of County Dublin in 1904, much of the area now in the city was green and wooded. Today, with the increase in urban development — the siting of houses, schools, churches and factories — much of the green area is now under concrete but, to our joy, many plants of great interest survive, in old doorways, cracks in pavements, rights of way, disused ground and so on. These include two woodland plants, the wood avens, *Geum urbanum* and the woodruff, *Galium odoratum*, and also the sedge, *Carex ovalis*, a plant of eskers. We also have limestone grassland species, now surviving on railway embankments, such as *Dactylorhiza fuchsii*, the common spotted orchid,

Briza media, the quaking grass, and *Blackstonia perfoliata* or yellow-wort. The Oxford ragwort, *Senecio squalidus*, is an example of a plant which established itself in Dublin recently and is now found in all divisions of the inner city. Mention could also be made of the many garden escapes which are now ubiquitous, for instance the butterfly bush, *Buddleja davidii*, feverfew or *Tanacetum parthenium*, and the rosebay willow-herb, *Epilobium angustifolium*.

Since Colgan's time, many new aliens have arrived and continue to arrive, often taking over derelict sites: for example, the willow-herb, *Epilobium adenocaulon*, which was first noticed in the third year of field work for this project and is now found in all 14 divisions.

Mention must be made of and gratitude expressed to the many botanists down the years who have stimulated the present generation of Field Club members, whose work and interest have brought this Flora into being. Here we remember James P. Brunker, Nathaniel Colgan, Joseph Doyle, Kathleen King, Arthur Stelfox, Jane Thompson and generations of past members. The tradition has been continued by Howard Hudson and Maura Scannell who have provided so much good advice and guidance for this Flora.

I express, on behalf of my committee and the members of the Dublin Naturalists' Field Club, sincere thanks to the Royal Dublin Society for its co-operation and help in the publication of this work. I wish this Flora of Inner Dublin a successful launching and may it give many years of pleasure to those who use it.

Helen O'Reilly
Dublin
1983

Introduction

In 1979 a study of the flora of Dublin's inner city was suggested to the Dublin Naturalists' Field Club. As the suggestion came during a time of petrol shortage, the idea was particularly attractive. The committee considered that the educational and scientific benefits of such a study would be of value to a great many people and would contribute to the knowledge of the city of Dublin.

Situated within the grid of 013 on the ½inch O.S. map, the inner city, as defined for the purpose of the Flora, is that area enclosed by the Royal and Grand Canals to the north and south respectively. The western boundary is delimited by a railway line, the Phoenix Park and the now infilled Grand Canal Harbour. The Irish Sea forms the eastern boundary (see map at the front or the back of the Flora).

The area was divided into 14 zones (see map), two consisting of the Grand and Royal Canals and the other 12 of more or less equal size radiating from the Liffey, following the natural grid lines of the streets. Some attempt was made to define particular areas within the city and place them in the same zone. For example, Trinity College and St Stephen's Green occur in the same zone and the suburban part of the city at Cabra forms a distinct zone. For the same reason the two canals comprise a single zone each.

At the outset of the work a field card was printed with the scientific names of the common city plants that would be encountered. The card was printed on one side only, the reverse side being for notes and additional species. When folded, the card is a convenient pocket size. Maps of the city were duplicated with zone boundaries clearly marked.

A team of about a dozen botanists began the field work, each being assigned a particular zone for the co-ordination of

1

records. Herbarium material (i.e. pressed, dried specimens) was collected so that a reference collection of critical and other plants would be available for reference. Two members of the Club subsequently co-ordinated all the records, organised further field excursions and collected the herbarium specimens.

During the three years of the work, field trips were held from May to September. These were official Club outings organised to record the flora of particular areas, and much field work was also carried out by individuals or small groups. In this way every region of the city was visited at least once during the course of the three years of the study. At the end of each field season the records for that year were transferred to individual species cards, thus showing the distribution of each species in the city by zone. It was therefore possible to identify which areas were in the greatest need of further study the following year.

A number of problems arose during the work which were tackled in the following ways:

1. *Critical Groups* — the treatment of critical groups presented a problem as such genera as *Taraxacum, Rubus* and *Fumaria* are difficult for experts to identify let alone a largely amateur Field Club membership. No attempt was made to define species of *Rubus* (except the easily recognisable *R. caesius* and *R. ulmifolius*). A representative collection of *Taraxacum* specimens was made in the city for determination by an expert, giving some idea of the forms that occurred. Other groups such as *Fumaria, Rumex, Epilobium* and *Polygonum* and genera of the Cruciferae were more manageable and specimens were collected and identified in the laboratory whenever necessary. With experience most of these genera could be identified with confidence in the field.

2. *Casuals and Aliens* — a high proportion of the plants encountered in the city were plants that are not native Irish plants (approx. 35%). A decision had to be reached about how to treat these species in the flora. It was decided to include every plant found in the city whether it be a casual, garden escape or other introduction, the only condition being that it should not have been planted deliberately in

2

that locality. For example in a waste ground, seedlings of the Turkey Oak, *Quercus cerris*, were found and included. It was not an ideal solution to the problem of what to include and what to omit but it was found to be the only workable solution. In some cases it was difficult to determine whether a plant had been planted or had become naturalised. For example in a number of waste places 'Virginia Creeper', *Parthenocissus* spp. were found growing on the ground. It was impossible to decide whether they were survivors of an old vanished garden wall or had become established from discarded plants.

3. *Inaccessible areas* — in the city there tend to be many sites which are inaccessible to the public, i.e. boarded-up waste grounds and back gardens. No doubt some plant species were missed because of this but generally it was found that most of the major waste ground sites became accessible at some stage or other. In a few instances hostile dogs made hazardous the recording of the flora in a particular site. Children were usually amused by the botanical activities and followed with interest, offering weeds to the botanists.

The city environment is in a constant state of change with waste grounds being cleared and built upon and others becoming vacant. Some sites recorded during the first and second year of the work were cleared of vegetation at a later stage. This causes a rapid change in the flora from year to year but was not of great concern to us. This Flora should simply be regarded as a glimpse in time of the city environment which we recognise might change radically within a few years. We hope that the Flora will therefore make a useful reference for future botanists in the city, who will be able to note the changes that have taken place in the composition of the flora from the time of study.

To the members of the D.N.F.C. the work proved to be a most interesting study. We hope it will provide a useful guide to the plants of the city for botanists and other interested people. It may also stimulate others to undertake similar work in other Irish cities and towns and provide comparative surveys in other urban parts of Ireland.

Peter Wyse Jackson

SENECIO SQUALIDUS

4

History of the flora

Declan Doogue

The first important botanical publication from Dublin's inner city did not appear in print until the opening of the eighteenth century, when Caleb Threlkeld's *Synopsis Stirpium Hibernicarum* (1726) was published. There are however a number of earlier plant records, mainly in the topographical literature, of various trees that had been planted in the environs of the city. Rows of flourishing limes were recorded towards the end of the previous century, in the grounds of Dublin Castle, and elms and sycamores were planted in the grounds of the Oxmantown Bowling Alley. Sycamore is now a well-established species throughout the city and elms still persist in several sites despite the ravages of Dutch elm disease and the many physical changes that have taken place within the city in the intervening three centuries. Political and economic conditions at the time could not have been conducive to good botanical field work but, despite this, several non-Irish naturalists visited the country in search of new and rare species. One of the first of these visitors was Richard Heaton, a military chaplain in the service of Thomas Wentworth (Walsh 1978). He discovered *Scilla verna* growing 'At the Ring's-End neere Dublin' some time between 1633 and 1641. Although parts of Ringsend come extremely close to the inner city area as defined for the present survey, it seems on balance better to exclude the record from the list of definite inner city plants. Heaton was followed to Dublin by William Sherard (who endowed the chair of Botany at Oxford in 1694) and later Edward Lhwyd (Keeper of the Ashmolean museum at Oxford) both of whom have botanical genera named in their honour (*Sherardia* and *Lloydia*). All three spent some time in Dublin, but it seems that none of their botanical observations has survived. While in Dublin, however, both Sherard and

Lhwyd (the latter better known for his archaeological investigations) visited the distinguished Irish scientist, Thomas Molyneux (1661-1733), and it is he who has the honour of being the recorder of the first flowering plant for the inner city. He noted, about the time of Lhwyd's visits in 1699 or 1700, a species of pearlwort which was found growing 'upon the Cape of the Wall of St. Mary's Churchyard' (Molyneux 1726). He named the plant '*Saxifraga graminea pusilla flore parvo tetraptelo*' but some doubt attaches to the exact species intended. Botanical nomenclature at the time was in its infancy and usage was by no means uniform. The cumbersome polynomial employed by Molyneux serves as a typical example. Colgan, in his *Flora of the County Dublin* (1904) assigns the record to *Sagina procumbens*, without comment, but it could also refer to *S. apetala*. (In Ray's *Flora of Cambridgeshire* for example, the name used for *S. procumbens* is '*Saxifraga Anglica Occidentalum*', while the name used for *S. apetala* is almost identical with that used by Molyneux for his Dublin plant.) Unfortunately, no relevant herbarium material has survived, and so it is not possible to resolve the matter further. Molyneux did acquire a *Hortus siccus* (a reference collection of dried herbs) in the course of his medical studies in Leiden, but this contains no Irish specimens (it is now preserved in the herbarium of the National Botanic Gardens) (Scannell 1979a). Today both pearlworts are widespread in Dublin, *Sagina procumbens* being commonest on pathways where it appears to be able to survive the severest trampling, while *S. apetala* does better in less stressful microhabitats on top and at the base of walls.

Thomas Molyneux was to have a profound influence on the whole process of scientific enlightenment in Ireland. He belonged to a family of philosophers, scientists and statesmen and pursued a highly successful medical career. He became State Physician, was conferred with a baronetcy in 1730 and was a prominent member of many of the scientific societies of the time, in particular the Dublin Philosophical Society which maintained a botanic garden and laboratory in Crow Street, off Dame Street (Hoppen 1970).

Being a respected public person he was asked to sit on

various committees and became one of the trustees of the committee charged with the construction of Steevens' Hospital, a position also held by Jonathan Swift, who was to satirise him in two of his less immortal poems, 'Mad Mullinix and Timothy' and 'Tom Mullinix and Dick'. Swift had a very poor opinion of science and scientists to judge by his descriptions of the Grand Academy of Ladoga in the third book of *Gulliver's Travels*. *Gulliver's Travels* was published on Friday the 28th of October, 1726, the day after Swift's lesser known neighbour, Caleb Threlkeld, is reputed to have published his *Synopsis*. To this work Molyneux contributed an appendix that included the enigmatic *Sagina* 'from St. Mary's Churchyard'. Threlkeld lived in Mark's Alley (now Mark's Alley, West, off Francis Street), little more than a stone's throw from St Patrick's Cathedral. Like Swift, he was an independent-minded clergyman who had many interests beyond his spiritual duties. At first he practised as minister to both souls and bodies but, finding greater success with the latter, he concentrated his efforts on physical healing and the related study of medicinal herbs. He describes his book as 'A Short Treatise of Native Plants, especially such as grow spontaneously in the Vicinity of Dublin; with their Latin, English and Irish names; And an Abridgment of their Vertues' but it is much more than that. The modern reader may find it bizarre, absurd, even vulgar in parts, but it is always entertaining and furthermore there is little reason to doubt the botanical accuracy of the information contained within its pages. The principal relevance of the *Synopsis* to this work is in the list of plants which are recorded as growing in or about the city. Most of the data contained in the list are clearly the work of Threlkeld himself, although it has been recently shown that much of the background information has been culled from other sources (Mitchell 1974). Many of the plants he lists growing in the city are still with us, often growing more or less where he found them more than 260 years ago.

On 'Martij 30, 1724', for instance, he found '*Alsine hederacea, sive Haederylae Folij, Morsus Gallinae Folio Haederulae,* Ivy Chickweed, or small Henbit "both flowering and fruiting" in the Hedge of a Garden at Stony Batter', This

7

plant, better known today as *Veronica hederifolia* L., still grows nearby in the grounds of the Incorporated Law Society, where it is represented by the subspecies *lucorum* (Klett & Richt.) D. Hartl. (*V. sublobata* auct.), the form more commonly encountered in hedgerows and woodland margins. In the course of the present survey a number of species more typical of shady habitats were found north of Stoneybatter. *Geum urbanum* (wood avens), *Geranium robertianum* (herb robert) and *Galium odoratum* (woodruff) may well indicate that the ancient wood (or at least fragments thereof) of Arbour Hill and the nearby Sallcock's Wood may have survived till a comparatively recent date. (The Coat-of-Arms of the Original Blue Coat School — King's Hospital — is said to have been carved from an oak that grew in this area.) Growing nearby, Threlkeld also noted '*Clematis Daphnoides Major*' which most likely refers to the great periwinkle, *Vinca major*, a well-known garden plant that frequently becomes established in the wild. Further north, he found growing 'upon the Mud Walls of Cabera-lane' the plant that he knew as '*Convolvulus Minor Vulgaris*', the small bindweed (*Convolvulus arvensis*). Stoneybatter and Cabera-lane (now Prussia Street and Old Cabra Road) formed part of the original Sligh Cualann, one of the great five roads of ancient Ireland, which ran from Glendalough to Tara. The mud walls seem to have disappeared but the small bindweed still grows on the lime-rich banks of the Royal Canal at nearby Phibsborough.

He saw *Anagallis arvensis*, the scarlet pimpernel, 'In fields and Sandy banks near the High-way to Drumcondra' and observed that 'It is reputed a Wound Herb, and is commended against Gripes of new born children, Falling sickness and Madness; The flower is gay and pretty to the eye, tho' it is a trailing plant'. At the World's End Tavern he found henbane, (*Hyoscyamus niger*). World's End Lane was later to be known as Mabbot Street, but little of the original district remains following the clearance of 'Monto' in the early twentieth century. Henbane is now an extremely rare plant, generally encountered only in coastal situations.

On the south side of the city he came upon a number of species that have become quite uncommon of late. *Chrysan-*

8

themum segetum, the corn marigold, grew 'in a muddy Bank of a Ditch in Patrick's-well-lane' — possibly that of the College Millstream at the end of Nassau Street. Nearer home he saw *Saxifraga tridactylites*, the rue-leaved saxifrage, 'upon Mr Grosvenor's Malt-house and some houses in Cavan-street' and *Descurainia sophia* (flixweed) growing 'among Rubbish, and upon some of the low Thatched Cabbins at the end of New-Street, near Black-pits'. *Descurainia sophia* is a declining species, now known in the county from only a handful of sites on the North Dublin coast. In England it was said to grow at that time only in 'filthie obscure base places' and this may explain its present scarcity, though certain botanists with an intimate knowledge of the inner city flora might, with good reason, assert the direct opposite.

One conspicuous member of the inner city flora that has shown no such decline is the grass *Hordeum murinum* which was noted by Threlkeld 'Upon the Sides of the Highway and on the walls leading to Bagatrath' (i.e. Baggot Street). It is now widespread in lanes, at the base of walls and on waste ground throughout the city. Its fruiting heads are used by Dublin children as 'arrows' or 'darts' — a practice which may have assisted the dispersal of the species to many parts of the city and beyond.

A little to the south of Black Pits was the area 'known as Roper's Rest' which was situated near the present day Greenville Avenue. Roper's Rest was at the most a ten minute walk from Threlkeld's home and he seems to have often botanised here in what was at the time the edge of the countryside. He records a number of plants from the general district, but many of these may be beyond the inner city limits. However, he says of the lesser celandine 'I have seen this Flower in March 27. 1726. under the Hedges between Roper's Rest and Dolphins-Barn'.

Almost all Threlkeld's botanical references relate to flowering plants. Bryophytes, algae and fungi are scarcely mentioned, but one entry is worth repeating saying, as it does, a great deal about Threlkeld, the times, and the state of contemporary knowledge of cryptogamic botany: '*Muscus innatus cranio humano*; Moss growing on a dead man's skull.

9

Frequent in Ireland where the poor people who are naturally hospitable, being misled by restless companions, rush into war, foolishly thinking to throw off the Blessing of the English Government. I took some from skulls upon the Custom-house-Quay, imported in large butts from Aghrim'. Moss that grew on old bones was said to have important medicinal properties, but further comment would not be helpful or particularly enlightening.

Very little is known of Threlkeld's life other than what can be gathered from the text itself. We know that he came from Cumberland, spent about 13 years in Ireland before the publication of his only known work and died two years later (Pulteney 1777). Richard Pulteney, who was a well-known naturalist of the day, and propagandist for the Linnaean system of classification, purchased a copy of the *Synopsis* some years after his death and discovered that, in it, a previous owner had inscribed supplementary details of his life and reported that he died 'after a short sickness of a violent fever at his house in Mark's Alley, Francis St. Ap. 28, 1728: and was buried in the new burial ground belonging to St. Patrick's near Cavan St.'.

Threlkeld had come to Dublin at a time when the city was just beginning to extend beyond its medieval limits. Urban development (mainly in an easterly direction) was to progress at a comparatively rapid rate from then on and was soon to influence and significantly modify the flora. Within a short period the Liffey was successfully embanked. Sir John Rogerson's Quay had been constructed, as had the inner part of the North Wall. A campaign of land reclamation was soon under way as the ground behind both quays dried out. These tracts of land became known as the 'Lotts' and were to be of considerable interest to botanists in later years.

The next botanist to make a contribution to our knowledge of the inner city was Walter Wade. Others, notably Isaac Butler and John Rutty, had been working in the interim on the flora of the county of Dublin, but they do not appear to have recorded anything from the inner city itself. Wade, like Molyneux, Threlkeld and Rutty, was a man of medicine and one of his early records was *Fallopia convolvulus* (black bindweed) from the precincts of Steevens' Hospital. Later, Wade became first curator of the Dublin Society's Botanic Garden at Glasnevin. Botany was now becoming recognised as a study in its own right and not just as a necessary part of the medical course. Wade's *Catalogus systematicus Plantarum indigenarum in Comitatu Dublinensi inventarum* (1794) was the first major Irish botanical work to employ a binomial nomenclature and a number of sites are mentioned for certain interesting plants. He found the sea purslane, *Halimione portulacoides* (which he called *Atriplex portulacoides*), 'in aggeribus vetustis juxta South Cumberland-Street in civitate Dublin' [on the old mounds near South Cumberland Street in the city of Dublin]. The sea purslane and indeed the ancient mounds and ramparts have long since gone, although South Cumberland Street lingers on, but a number of maritime species still live in some very unlikely ground nearby. *Spergularia rupicola* grows on an old wall between Trinity College and the Liffey and *Desmazeria marina* occurs occasionally on derelict sites nearer the river.

While Wade was working on his *Catalogue* the North and South Circular Roads were under construction. He recorded *Galium palustre* (marsh bedstraw) 'In locis humidis et ad rivulos, apud Long Meadows, Phoenix Park, Circular-Road et alibi' [in moist places and by streams, at Long Meadows, Phoenix Park, Circular-Road and elsewhere] and *Calystegia sepium* 'in sepibus fere omnibus abunde, sed praecipue apud Rathmines et Circular-Road' [in abundance in pretty well all hedges, but particularly at Rathmines and Circular Road]. He also found *Saxifraga tridactylites* (rue-leaved saxifrage) 'super tecta et muros antiquos praesertim lutulenta juxta Lucan-turnpike, in Long-lane inter Camden-Street et New-Street' [over roofs and old walls, particularly muddy ones near Lucan turnpike, in Long-lane between Camden Street and New Street] close to the original Threlkeld localities.

Ten years later Wade published his *Plantae Rariores in Hibernia Inventae*, which included a few possible inner city plants from the vicinity of the North Walls (*Lepidium ruderale* (narrow-leaved pepperwort) — a casual and *Ruppia maritima* (a pondweed) 'in the salt water marshes', but he also found

11

coriander (*Coriandrum sativum*) 'on the ground to the right from the upper end of Townsend-street to the draw-bridge going to Ringsend'.

In 1818 the *History of the City of Dublin* was published. It was written by John Warburton, Rev. James Whitelaw and Rev. Robert Walsh (Warburton *et al.* 1818). Walsh contributed an appendix (no. XIII) to volume 11 of this extensive work, which lists the plants found along the Dublin coast. Of soapwort, *Saponaria officinalis*, he says 'This plant grows on the North Strand near Ballybough Bridge'.

Wade's counterpart in the Trinity College Botanic Gardens was James Townsend Mackay. The gardens were at that time situated on Lansdowne Road where the Berkeley Court Hotel now stands. Mackay prepared a *Catalogue of the indigenous plants of Ireland* which was published in the Transactions of the Royal Irish Academy in 1825 and incorporated a number of botanical records published 20 years earlier (Mackay 1806). Unfortunately, a large number of his Dublin discoveries were made on the south side of the Grand Canal in the immediate vicinity of the College Gardens and were clearly growing outside the inner city margins. However, he was quick to appreciate the botanical potential of the canal and found a number of aquatic species growing in its waters. He recorded a pondweed as *Potamogeton pusillus* (*P. berchtoldii*?) 'In the Grand Canal above the first lock' and *Sparganium erectum* in a ditch by its banks between Macartney Bridge (i.e. Baggot Street Bridge) and Donnybrook Bridge (Leeson Street Bridge), close to his home (at 5, Haddington Terrace) and place of work. On low wet ground between Mark's Church and Ringsend he discovered marsh yellow-cress or *Rorippa palustris* (as *Nasturtium terreste*) and found blue fleabane (*Erigeron acer*) on the outer side of the North Wall below the new (Gandon's) Custom House. He also noted nettle-leaved goosefoot (*Chenopodium murale*) behind the old Anatomy-House, Trinity College, and recorded (possibly in error) another goosefoot *Chenopodium olidum* between the Custom House and Annesley Bridge. On the banks of the nearby Royal Canal he found the sedge, *Carex pendula* 'between Dublin and Glasnevin'.

12

The construction of the canals had a major enriching influence on the flora of the inner city. Work on the excavation of the Grand had begun as early as 1755 and on the Royal shortly after. Both canals drew water from a variety of sources and in due course aquatic plants (and animals) began to spread from the Shannon basin and the lakes of the central plain, assisted by the dragging action of the barges and the downward flow of the water towards the city.

The barges that first worked the canals were horse-drawn and motorised boats were to follow soon afterwards. For a while both canals prospered, but with the advent of the railways, the volume of traffic began to decline.

Mackay was not the only one to appreciate the value of the canal system to botanists. In 1833 *The Irish Flora* appeared, published anonymously but written by Lady Kane (Katherine Sophia Baily) and incorporating many of the botanical records of James White, a gardener in the Glasnevin Botanic Gardens. Lady Kane was clearly impressed by the pioneering work of Wade and his successor Dr Samuel Litton in the popularisation of botanical studies, and is lavish in her praise of White's work. Of him she says 'For the past thirty years he has investigated with unremitting attention the botany of this country; and the intelligence and unwearied zeal with which he acquired his very extensive knowledge is only equalled by the liberal spirit with which he is always ready to communicate this information to any person who is at all desirous to become acquainted with any branch of the science' — a feature which happily still characterises the activity of many of the natural history organisations of the present day.

Unlike earlier works *The Irish Flora* provides descriptions as well as localities for each species. These are drawn from Smith's and Hooker's *English Flora* and *British Flora* and a certain amount of habitat information is also provided. Numerous plants are recorded from the banks of the Royal Canal and many of these are directly relevant to this Flora. Hemlock (*Conium maculatum*) grew 'along the banks of the Canal leading by the Lots', purple loosetrife (*Lythrum salicaria*) was 'along the canal side from Dublin to Glasnevin',

13

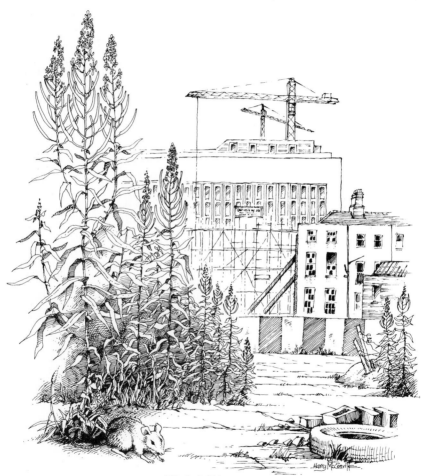

EPILOBIUM ANGUSTIFOLIUM.

14

and from the North Strand there was an apparent confirmation of the *Saponaria* record made by Walsh 15 years earlier, 'Var. with double flowers, grows on the top of a low wall near the Sunday School, North Strand'.

Also recorded at Cross Guns Bridge (and further west) were tufted hair-grass (*Deschampsia caespitosa*), marsh arrow-grass (*Triglochin palustris*), water plantain (*Alisma plantago-aquatica*) and water milfoil (*Myriophyllum verticillatum*). The last-mentioned plant was also found in the old City Basin (or Bason) near James' Street, which had served as a reservoir for part of the city water supply since 1722. In the eighteenth and nineteenth centuries the area was a fashionable promenade and when the basin was drained 'the bottom was immediately covered with the plant'. The pot herb, alexanders or *Smyrnium olusatrum*, grew nearby. The marsh marigold, *Caltha palustris*, was 'abundant in a low field near the Grand Canal Harbour'. The banks of the Grand Canal had also been examined and, on the stretch within the inner city limits west of Portobello, purging flax (*Linum catharticum*) and the sedges, *Carex remota* and *C. pendula*, were found.

One of the most interesting and significant records in *The Irish Flora* is that of lesser swine-cress or *Coronopus squamatus* which was seen 'in the little road leading from Sinnot-place to the Circular Road, covering the ground abundantly'. The plant at the present time does not actually grow in the lane-way but it was found in small quantity in 1981 in waste ground at the end of Synnott Place near Dorset Street. Unfortunately the site was developed shortly after but the species may still survive, possibly as dormant seed, in some waste ground nearby. Winter-cress or *Barbarea vulgaris* was found near the North Circular Road 'along the ditches from Phibs-borough Chapel to the Park Gate', and deadly nightshade, *Atropa belladonna* grew 'on the side of a ditch at the Circus between Blessington St. and the Penitentiary'. This very poisonous plant seems to be gone from the inner city, but has been seen recently at Killester and on Pembroke Road (where it may have escaped from the old Trinity Botanic Gardens). Another declining Irish plant, penny cress or *Thlaspi arvense*, was found at Black Pits, and the grass *Parapholis strigosa*

15

found 'along the roadside below the Custom House Docks' may well scrape into the inner city area, indicating that the maritime influence had not been entirely lost as the North Lotts continued to dry out. Other plants had also been recorded from the Lotts but the area was quite extensive and it is impossible to precisely locate most of these.

Mackay in the meantime had not been idle. In 1836 he published his *Flora Hibernica* in which the plants are arranged according to the 'Natural System'. Previous works, although employing a binomial nomenclature, were based on a convenient but archaic system of classification and the 'Natural System' took into consideration the obvious relationships that existed between certain genera and families of flowering plants. Although this work contains few localities for plants, the scurvy grass, *Cochlearia danica*, was recorded below the Dublin Custom House. Mackay died in 1862 but one of his assistants at the Trinity College Botanic Gardens, David Moore (who later became curator of the Botanic Gardens at Glasnevin), was approached by Alexander G. More with a view to publishing a *Cybele Hibernica* along the lines of H.C. Watson's *Cybele Brittanica*, which had appeared in 1860. This work was to be the first serious attempt to outline the geographical distribution of the various flowering plants and ferns of Ireland. Despite the magnitude of the undertaking, compilation proceeded rapidly and included a few additional records for the inner city, all from the canals.

The invasive Canadian pondweed, *Elodea canadensis*, a native of North America, had appeared some years previously in Dublin and in *Cybele Hibernica* it was noted that it had become abundant in the canals near Dublin. It was to become a serious pest, often choking the waterway, although nowadays it is far less troublesome. David Moore noticed the much rarer *Groenlandia densa* in the Grand Canal (where it still grows) near Portobello. It is now one of the species protected by law in Ireland because, although it is common in this country, it has become extremely rare in Europe as a whole because of water pollution. Moore considered that it had been brought to Dublin by turf boats from the midlands (Moore and More 1866).

16

The other collaborator in the *Cybele Hibernica* work, A.G. More, discovered the horsetail, *Equisetum variegatum* (as *E. wilsonii*) growing sparingly a little below Cross Guns Bridge, Glasnevin. This species is abundant by the Royal Canal from Clonsilla westwards, and was certainly at Cross Guns Bridge up to about 1970, but has not been found since despite careful search.

Some years after the publication of *Cybele Hibernica*, A.G. More produced a supplement entitled *On Recent Additions to the Flora of Ireland* (1872). It adds virtually nothing to our knowledge of the inner city flora, but corrects some errors of identification made by earlier authors. However More noted the appearance of *Coronopus didymus* 'near the canal at Balls Bridge'. *Coronopus squamatus* was till then the more frequent species, but now *C. didymus* has become by far the commoner. *Recent Additions* also includes a record contributed by M. Dowd of the College Botanic Gardens. He discovered *Berula erecta* (*Sium erectum*) or narrow-leaved water parsnip in the Canal, near the North Strand, Dublin. Dowd botanised extensively in the Midlands and was to contribute many records for Leinster in the supplement.

By now, the late Victorian interest in natural history was in full swing and many authors prepared papers of national and local importance, many of which appeared in the pages of *The Irish Naturalist* -- a monthly journal which was first published in April 1882. In the November 1896 issue the Rev. Thomas Gibson reported on 'The Botany of a School Playground in the Heart of Dublin' (Gibson 1896). In the course of the previous 18 years he had noted a wide variety of flowering plants growing in the grounds of the King's Hospital School, situated between Blackhall Place and the Royal (Now Collins') Barracks. Unfortunately Gibson had an infuriating habit of collecting seeds of wild plants during his summer holidays and scattering them about the school yard on his return. Modern botanists would shudder at such a practice as it becomes extremely difficult in some cases to decide exactly which species are planted or introduced and which occur more or less naturally. He admits to introducing certain plants such as yellow horned poppy (*Glaucium*

17

flavum), green hellebore (*Helleborus viridis*) and bluebells (*Hyacinthoides non-scriptus*), and records both periwinkles (*Vinca major* and *V. minor*), one of which had been reported by Threlkeld 170 years earlier. However if one leaves aside certain very obvious introductions and also disregards the species that are widespread throughout the country, there remain a number of plants which are indicative of distinct habitats. Most significantly (recalling the observations made earlier regarding the survival of a woodland flora) is the presence of a small group of shade-loving species, such as *Glechoma hederacea* (ground ivy) and *Circaea lutetiana* (enchanter's nightshade) as well as *Hypericum androsaemum*, or tutsan (possibly planted), and *Galium odoratum* (woodruff), which, as mentioned above, still grows nearby.

There is, additionally, a group of plants most frequently considered to be characteristic of dry calcareous grassland. These include bulbous buttercup *Ranunculus bulbosus*, purging flax *Linum catharticum*, dove's-foot cranesbill *Geranium molle*, wild carrot *Daucus carota*, bird's-foot trefoil *Lotus corniculatus* and lady's bedstraw *Galium verum* — an assemblage which still occurs less than a quarter of a mile away outside the inner city boundary, near the Magazine Fort in Phoenix Park. The most sizeable proportion of his list, however, is the group of plants that have become inner city specialists. These are species that are seldom encountered in Ireland far from habitation. Some are genuine garden escapes like ivy-leaved toadflax (*Cymbalaria muralis*) and evening primrose (*Oenothera biennis*) which have become established on walls and in waste places. Others are typical weed species, like fool's parsley *Aethusa cynapium*, scentless mayweed *Matricaria perforata*, sun spurge *Euphorbia helioscopia* and annual mercury *Mercurialis annua*. Gibson recorded in excess of 130 species from the school grounds, a number that could have been considerably increased had he included the grasses, ferns and other species with which he was not familiar. Examining his list, one gets the impression of a naturalist who lacked a detailed knowledge of the Irish flora, but who was aware of his limitations and, rather than run the risk of recording certain species in error, chose wisely to concentrate on plants that were relatively easy to identify.

In 1895 the death occurred of A.G. More. Two of his protégés, Nathaniel Colgan and Reginald Scully, were given the task of updating the *Cybele Hibernica* and a second edition appeared in 1898. Colgan had been quietly working on a projected *Flora of the County Dublin* (Colgan 1904) for some years and this appeared in 1904. It is an exceptionally comprehensive piece of work and is still considered to be one of the best county floras ever written in this country. He visited the entire county during his investigations and recorded a substantial number of plants from the inner city. Colgan was well aware of the enriching effect that the canals were having on the flora of the Dublin district. He visited the Royal on several occasions and recorded such aquatics as *Ranunculus circinatus* (a water buttercup), hemlock water dropwort *Oenanthe crocata*, gipsywort *Lycopus europaeus* and arrow-head or *Sagittaria sagittifolia*. Although *Oenanthe* and *Lycopus* are still conspicuous along the margins of the canal, the true aquatic species have declined noticeably in the inner city stretch. *Ranunculus circinatus* was not re-found in the course of the present survey and an old site where it once grew (Blacquiere Bridge) is also lost. It had been collected there by Richard Harrington in 1859 but the stretch of canal that linked the Broadstone Station with the main canal has long been filled in. Colgan also botanised on drier ground nearby. He found hoary cress *Cardaria draba*, pineapple-weed *Chamomilla suaveolens* and dwarf mallow *Malva neglecta* in the vicinity of Cross Guns Bridge, and *Cymbalaria muralis* ivy-leaved toad flax, annual mercury *Mercurialis annua* and the grass *Bromus erectus* nearer Broadstone Station. *Malva neglecta* still survives at Cross Guns Bridge and *Chamomilla suaveolens* has become established on roadsides and waste ground throughout the country. Colgan first recorded this species for Ireland in 1894 from the Carrickmines district.

A species that received considerable botanical attention due to its dramatic decline was *Sisymbrium irio* (London rocket). It was believed to have been widespread about the city although specific inner city localities are lacking. Colgan noted that it had of late become quite rare but recorded it at Clanbrassil Street, Mount Brown and Marrowbone Lane. He attributed its decline to the introduction of concreted foot pavements replacing the previously unpaved footways at the base of walls, but other factors may have been involved. He also examined the flora of the Grand Canal but added little on the inner city stretch. He found *Potamogeton pectinatus* and *Zannichellia palustris* (pondweeds) there and recorded the sedge, *Carex pendula* at Leeson Street Bridge, a little to the east of the sites mentioned by Lady Kane.

Colgan published little from then on, concerning himself more with marine Mollusca and Tunicata than botany. His Flora, however, is valuable not only for what it includes but also for what it does not mention. Due to the thoroughness of his field work, it is possible to say that certain species were almost certainly unknown in the inner city up to 1904. Yellow water lily (*Nuphar lutea*) for instance was not mentioned, nor was Oxford ragwort (*Senecio squalidus*) or eastern rocket (*Sisymbrium orientale*) all of which grew only on the periphery of the inner city in Colgan's time.

In 1918, Scully met J.P. Brunker, who was later to emulate Colgan's work with his *Flora of County Wicklow* (Brunker 1950). Scully, recalling the assistance given to him by A.G. More, set about initiating Brunker 'into the living tradition of field work' — a tradition that has been carried on through the membership of the Dublin Naturalists' Field Club to the present day. Brunker was employed at the Guinness Brewery and took the opportunity to catalogue the flora of the immediate area. He recorded 200 species and prepared a report entitled the *Flora of St. James's Gate*, dated 27th Nov. '44 (now in the possession of H.J. Hudson). Many of the species listed are ephemerals and several are exotics, present only as contaminants in foreign grain brought to the brewery. Also included are a number of aquatics found in the Grand Canal Basin, as well as a number of native species that had managed to establish themselves in the precincts of the Brewery. *Ranunculus circinatus* a water buttercup, brooklime (*Veronica beccabunga*), gipsywort (*Lycopus europaeus*), ivy-leaved duckweed (*Lemna trisulca*), water plantain (*Alisma plantago-aquatica*) and a number of *Potamogeton* species were found in the Canal Basin. Despite the decline in canal traffic, the brewery line had remained functional and turf boats transporting fuel to the capital during the 'Emergency' (1939-1945) had, if anything, stimulated the flow of hydrophytes into the city. Among the grain aliens, tall rocket (*Sisymbrium altissimum*) and eastern rocket (*S. orientale*) were recorded as was field penny cress (*Thlaspi arvense*) and tall melilot (*Melilotus altissima*). The report also gives important pre-1944 dates for the swine cress *Coronopus didymus* (believed to have been introduced in poultry feed), the rosebay willow-herb *Epilobium angustifolium, E. roseum* and Oxford ragwort (*Senecio squalidus*), as well as many other species that were to become permanent members of the inner city flora. It is not possible to do justice to this report in a brief overview, but one other group of plants encountered must be mentioned.

On the Liffey wall just outside the brewery at Victoria Quay, Brunker recorded a number of native species that appear to have made their way, unassisted by man, to the city. Meadowsweet (*Filipendula ulmaria*), water valerian (*Valeriana officinalis*) hemp agrimony (*Eupatorium cannabinum*), water mint (*Mentha aquatica*) and wall pellitory (*Parietaria diffusa*) all grew here, and most still do. He also recorded *Sagina ciliata* on the railway by the 'Painter's Shop, Victoria Quay'. *S. ciliata* is now regarded as being a subspecies of *S. apetala* but it can still be distinguished from the more common form of *S. apetala* that is widespread throughout the city.

The Grand Canal Basin was filled within the last decade and in it the large flax, *Linum usitatissimum* was found. It is an alien species seldom encountered in Ireland, but it too is on Brunker's list, made almost half a century earlier, which is of interest.

Ireland's most distinguished botanist, Robert Lloyd Praeger, worked in the National Library in Kildare Street and lived first in Rathgar and later in life at 19 Fitzwilliam Square. In 1901 he published *Irish Topographical Botany* (Praeger 1901) which summarised the geographical distribution of the Irish flowering plants as they were known then. Twenty years later, he contributed to knowledge of the inner city flora with a short list of ferns published in *The Irish Naturalist* (Praeger 1920), recording hart's tongue fern (*Phyllitis scolopendrium*), male fern (*Dryopteris filix-mas*), broad buckler fern (*D. dilatata*) and the soft shield fern (*Polystichum setiferum*) over the pediment at the southern end of Leinster House, and found bracken (*Pteridium aquilinum*) in nearby Dawson Street. He also noted that *Dryopteris filix-mas* had until recently also grown at the north end of Harcourt Street. In *The Way that I Went* (Praeger 1937) he records the demise of the four species at Leinster House (the seat of Dáil Eireann) following clean-up operations but, by way of compensation, he announced the presence of the lady fern (*Athyrium filix-femina*) and hart's tongue fern (*Phyllitis scolopendrium*) near his home in Fitzwilliam Square which is within the inner city.

From 1935, the Dublin Naturalists' Field Club, of which Praeger was a prominent member, had been engaged in a long-term revision of Colgan's *Flora* and in 1961 a Supplement to that work was published (Dublin Naturalists' Field Club 1961). Many of the inner city records are obviously drawn from Brunker's observations, but a small number of additional records are included. Nodding bur-marigold (*Bidens cernua*) was found in 1922 at Portobello Bridge, corn chamomile (*Anthemis arvensis*) in 1924 at the Custom House Quay, Oxford ragwort (*Senecio squalidus*) in Taylor's Lane, Pimlico in 1935 and further records of wormwood (*Artemisia absinthium*) from High Street and Francis Street. Strangely, the rosebay willow-herb, *Epilobium angustifolium* had still not been recorded for the inner city, although the compilers of the Supplement had noted it from peripheral locations.

The hornwort, *Ceratophyllum demersum*, was found in the Grand Canal at Huband Bridge in 1970, illustrating the still dynamic nature of the canal flora (Scannell 1971). It has since been found at several points along the canal and must be considered a recent arrival. In 1972, D.A. Webb recorded both *E. angustifolium* and *Senecio squalidus* from the perimeter wall on the north side of Trinity College and additional

22

localities for the latter species on Essex Quay and Killarney Street (Webb 1972). Six years later M.J.P. Scannell noted *E. angustifolium* in nearby Fownes Street (Scannell 1979b). It is remarkable that such an aggressive species should have taken so long to become established (or at least recorded) in the inner city. It may well be that the successful establishment of species such as these may depend more on real estate prices and the availability of planning permission than on biogeographical chance!

The development of our knowledge of the flora of the inner city is paralleled by the physical development of the urban habitat itself. Many of the early botanical records suggest an environment where the juxtaposition of rural and urban habitats provided a mixture of native and ruderal species that modern botanists would seldom expect to find in close proximity. Many of the weed species have been introduced inadvertently into the city both as impurities in animal feed and as contaminants in grain, being brought in open carts to the many small breweries and mills that flourished in Dublin up to the opening of the present century. The introduction of bulk grain transporters and containerised traffic since the 1950s has no doubt cut off an interesting supply-line of unusual species. This, coupled with the virtual disappearance of horses and cattle, has diminished the rural element in the city flora. Until quite recently cattle were brought to market and driven cowboy-style down the North Circular Road on their way to the North Wall for export.

Floristic studies as in this Flora are enjoying somewhat of a revival at the present time, and opportunities for botanical investigation are open to everyone. This is in marked contrast to the conditions under which the early botanists were obliged to operate, as Threlkeld saw it in 1726: 'The only Reasons I know why this Branch of Learning has been dormant in Ireland, and no publick Advances made towards its Illustration, are that the Wars and Commotions have laid an Imbargo upon the pens of the learned, or discord among the petty subaltern Princes has render'd Perambulation perillous ... The Nation being now calm and settled under a wise and just King, every one may follow the Muses'.

Perhaps, on reflection, things have not changed all that much.

Some sites of botanical interest

Micheline Sheehy Skeffington and **Peter Wyse Jackson**

Introduction

If the Flora is to be used as a guide to the inner city for naturalists, it is useful to note localities where a good variety of species can be found. Also, as the Flora is essentially an account of plant assemblages as they existed during the survey years (1979-1981), it will be of interest to see the changes that may take place in future years in particular localities.

Due to the transient nature of the inner city environment it is difficult to select sites which are likely to remain for some time and can therefore be of botanical interest in the years to come. There are however some areas which have never been built upon nor developed and hopefully will remain relatively unaltered. In addition others are worthy of mention because their flora is particularly interesting or diverse.

Sites

Railway Embankments: The northwest boundary of the inner city runs along a series of railway lines which provide a link with the adjacent countryside of County Dublin. Following the Royal Canal tow-path, west of Cross Guns Bridge, one soon reaches the railway line. The embankment is made up of a dry, well-drained limestone soil, providing a dry calcareous habitat very similar to the gravelly and limey esker ridges common in the midlands. The vegetation of these railway banks is therefore that of a calcareous grassland with some scrub and has very little in common with other inner city habitats.

Many of the species of grass are lime-loving (calcicole) and include *Briza media* (quaking grass), *Avenula pubescens* ('hairy oat') and *Trisetum flavescens* ('yellow oat'). Two legumes *Anthyllis vulneraria* (kidney vetch) and *Lotus cor-*

niculatus (bird's foot trefoil) are also common and three species of orchid were noted: *Anacamptis pyramidalis* (pyramidal orchid), *Dactylorhiza fuchsii* (common spotted orchid) and *Gymnadenia conopsea* (fragrant orchid). An interesting member of the flora is *Orobanche minor*, a broom-rape that grows parasitically on the roots of clovers.

A second stretch of railway embankment lies close to St Attracta Road and Fassaugh Avenue (zone 12). Its flora is not typically that of limestone grassland, but it is the only inner city site for *Blackstonia perfoliata* or yellow-wort, common on limestone soils and sand dunes in Ireland. *Lotus corniculatus* is again abundant there and the grassland species, *Hypochoeris radicata* (cat's ear), *Centaurea nigra* (blackheads) and *Vicia sepium* (a vetch) also occur.

These gravelly sites dry out quickly in summer and are best visited in spring or early summer when many of the species are in flower.

Royal Canal: This canal has been neglected for a long time and only now is receiving attention from various conservation groups. Hopefully it will soon be fully restored. Although it has not as rich an aquatic flora as the Grand Canal, its banks have nevertheless an abundant and characteristic flora. This includes the reed *Phalaris arundinacea* and the reed-like grass *Glyceria maxima*. *Iris pseudacorus* (the yellow flag iris), *Juncus inflexus* (a rush), *Filipendula ulmaria* (meadowsweet) and *Angelica sylvestris* (wild angelica) all grow in many places along the banks. Close by the edge are *Hippurus vulgaris* (mare's tail), *Nasturtium microphyllum*, *N. officinale* (water-cress) and the hybrid between these two. Other species common along the banks in wet places are *Lycopus europaeus* (gipsy-wort), *Mentha aquatica* (water-mint) and two umbelli-fers, *Apium nodiflorum* (fool's watercress) and the poisonous *Oenanthe crocata* (water-dropwort). Submerged species which are less accessible from the edge include *Elodea canadensis* (Canadian pond-weed) and *Nuphar lutea* (the yellow water lily). Trees are infrequent along the banks but ash, sycamore, alder and a willow (*Salix atrocinerea*) do occur.

The canal is bounded by steep walls for most of its inner

26

HARRY McCONVILLE.   MYRIOPHYLLUM SPICATUM

city course. These walls are a particularly favoured habitat for ferns such as the spleenworts and hart's tongue. Near where the canal meets the Liffey (zone 8) the banks have crumbled and a patch of marsh ground supports the salt marsh species, *Aster tripolium* (sea aster) and also the sedge *Carex otrubae* which is frequent in ditches by the sea. The walls at the junction of the canal and the Liffey have three interesting species, all of which have predominantly coastal distributions in the country, *Desmazeria marina* ('darnel poa'), *Lepidium latifolium* or dittander and *Parietaria diffusa* or wall pellitory.

Grand Canal: This canal and its banks are more frequently tended than those of the Royal Canal, its bottom is dredged and the vegetation of the verges is cut annually. The constant clearing of the grassy margins may help the advent of some unusual species, including *Pentaglottis sempervirens* (alkanet), *Aster novi-belgii*, an aster and the mint *Mentha suaveolens*, all garden plants. Planted and self-sown trees along the banks add to the habitat diversity by providing shade and also an input of nutrients to the canal from falling leaves.

The aquatic vegetation at the margins is similar to that of the Royal Canal, but the submerged or partially floating species are much greater in diversity. Within the inner city those species seen in the Grand Canal which were absent from the Royal include *Groenlandia densa* (the broad-leaved pondweed), *Ceratophyllum demersum* (hornwort) and the pondweeds *Potamogeton obtusifolius*, *P. crispus* and *P. fili-formis*. More conspicuous are the emergent plants, *Sagittaria sagittifolia* (the arrowhead) and *Alisma plantago-aquatica* (the great water plantain). Constant weed killing and dredging of this canal may in fact increase (or maintain) the diversity of the aquatic flora by reducing the growth of more vigorous species.

Two sites of special character along the canal should be noted:
1) at Grand Canal Quay (zone 1) where *Ulex europaeus* (gorse), *Juncus effusus* (common rush) and *Silene alba* (the white campion) are found. The presence here of two species

28

of duckweed, *Lemna minor* and *L. trisulca*, indicate that the waters are relatively undisturbed and slow flowing.

2) Some allotments still exist along the canal at Dolphin's Barn (zone 2) where a few ruderals of cultivated ground are found, including *Lamium purpureum* (the red dead nettle), *Spergula arvensis* (corn spurrey) and *Urtica urens* (annual nettle). Further along the canal the osier *Salix viminalis* has been planted and may have been formerly used for thatching and basket-making.

It is easy to walk along both banks of the canal for most of their length and it is fruitful ground for the botanist.

'Woodland' Localities: Some habitats in Dublin city seem to have been disturbed little since before the days of urbanisation. The grounds of large institutions often provide a habitat for plants more characteristic of the rural landscape. St Brendan's Hospital (zone 13) is an example of a locality in the inner city where the presence of certain species is thought to date back to times when woodland or hedgerows still existed nearby. Study of old maps would suggest that no building or major clearing has occurred on much of the Hospital site, and it is unlikely that many of the woodland and hedgerow plants found there would have been introduced subsequent to the building of the hospital. In the more shaded areas one may find *Glechoma hederacea* (ground ivy), *Hypericum androsaemum* (tutsan), *Hedera helix* (ivy), *Geranium robertianum* (herb robert) and *Stellaria holostea* (the greater stitchwort).

Another possible relict woodland site is in an old garden beside the former cattle market off Aughrim Street (zone 13). Here *Fraxinus excelsior* (ash), *Acer pseudoplatanus* (sycamore), as well as *Sambucus nigra* (elder) predominate. Ivy is abundant and *Solanum dulcamara* (bittersweet) is also present. Shade-tolerant herb species are represented by *Geranium robertianum* (herb robert), *Geum urbanum* (wood avens) and *Galium odoratum* (woodruff). It is unlikely that these species would have been deliberately introduced, though *Petasites fragrans* (winter heliotrope) may well have been. It prefers moist shady places and, though not native in Ireland, is popularly grown for its fragrant winter flowers.

Mixed Grassland: A large waste ground with a substantial variety of species is found by Montpelier Gardens off Infirmary Road (zone 13). It is now under threat of development in the next few years, but deserves special mention for its flora which include a number of hedgerow and grassland species. This site has certainly never been built upon in the past. The grass species represented are mainly coarse wasteground species such as *Arrhenatherum elatius* (false oat grass) and *Dactylis glomerata* (cock's foot grass), but there is also a variety of sedges and rushes to be found: *Carex hirta, C. ovalis* and *Juncus conglomeratus.* Other unusual species occur including *Odontites verna* ('red bartsia'), *Stachys sylvatica* (woundwort), *Eupatorium cannabinum* (hemp agrimony), *Geranium robertianum* (herb robert), *Primula veris* (cowslip), *Rosa canina* (the dog rose) and an eyebright, *Euphrasia sp.* The variety of species suggests the presence formerly of hedgerows and moist shady banks surrounding damp rushy fields.

Numerous waste ground species were also noted and many discards from cultivation such as potato, snapdragon, *Fuchsia,* lilac, virginia creeper and raspberry. This locality is perhaps the richest in number and diversity of plants comprising in all about 120 species. Sadly it may disappear in the next few years.

Walls: Many of the old and crumbling walls of the city are colonised by large populations of plants. They are a favoured habitat for ferns wherever there is sufficient moisture seeping down the wall. Common species seen are *Asplenium ruta-muraria* (wall rue), *Phyllitis scolopendrium* (hart's-tongue fern), *Dryopteris filix-mas* (male fern) and *Polypodium vulgare* (common polypody). Other plants common in waste grounds also colonise this habitat, thriving even in small pockets of soil and crumbling mortar. *Buddleja davidii* is a frequent sight, *Cymbalaria muralis* (the ivy-leaved toadflax) is common and *Senecio squalidus* (Oxford ragwort) is abundant on wall-tops. There should be no difficulty in locating such walls with a rich flora which can be found throughout the city.

Waste Grounds: The decaying centre of Dublin has a wealth of derelict sites scattered throughout most parts of the city. New ones arise almost daily as old buildings are demolished to make way for future development, and many subsequently lie derelict for years. During this time a rich wasteland flora develops often covering acres in extent. Others are smaller or less long-lived but even these often contain many plants of interest. One can observe a successional cycle in the flora on a waste ground from its beginnings until it is redeveloped. In the first year when the ground is disturbed and devoid of vegetation the first colonisers are, for the most part, annual weeds. The second year sees the establishment of more permanent plants on the soils with a higher fertility, annual weeds being out-competed and surviving only on rubble heaps and on the poorest soils. *Capsella bursa-pastoris* (shepherd's purse), *Hordeum murinum* ('wall barley'), *Stellaria media* (chickweed), *Polygonum aviculare* (knotgrass), *Fumaria species* (fumitories), *Epilobium species* (willow-herbs), *Poa annua* (annual meadow grass), *Sisymbrium officinale* and *S. orientale* (hedge mustards) are all common primary colonisers in the first year. In later years the number of perennial and native species increases and *Buddleja davidii*, *Acer pseudo-platanus* (sycamore), *Reynoutria japonica* (Japanese knotweed) are common. Clovers and perennial grasses arrive eventually forming a closed community not unlike rough pasture. Common in later years are *Senecio jacobaea* (ragwort), thistles (*Cirsium* species), *Potentilla anserina* (silverweed), *Ranunculus repens* (creeping buttercup) and the grasses *Lolium perenne* (perennial rye grass), *Poa pratensis* and *Poa trivialis* (meadow grasses) and *Arrhenatherum elatius* (false oat).

However, few waste grounds survive to reach a late stage in their development. It is interesting to speculate as to what would be the final stable or 'climax' vegetation of a Dublin waste ground. On the richer soils it would probably become rough grassland perhaps with elder (*Sambucus*) and, on the poorer shallow soils, a scrub-type vegetation of sycamore, *Buddleja*, willows and a few resiliant herbs such as the mayweeds, *Matricaria maritima* and *M. perforata*.

31

During the field work time of this study a fine waste ground existed at Charlemont Street (zone 4) where many of the stages in vegetational colonisation were represented at the same time in different parts of the site. At the time of writing much of this site is being developed by Dublin Corporation for housing and will soon disappear. Close by in Earlsfort Terrace a large wasteground stands on the site of the Old Alexandra College (zone 3), opposite the National Concert Hall. Here are the remains of an old garden, shaded by mature trees, some rough grassland and a large area of *Buddleja* and sycamore 'scrub'.

The area on the south side of the quays towards the mouth of the Liffey (zone 1) is designated for redevelopment and in the last few years a large area has been cleared of buildings. Plant colonisation is taking place rapidly and in a year or so this area may yield interesting plants for the botanist. Close proximity to the port of Dublin would help the establishment of interesting aliens brought by ships and their cargoes.

The north side of the inner city has many waste grounds but few are as large in extent as those mentioned above. The area along the banks of the Royal Canal near Sheriff Street (zone 8) was the single largest site where a typical ruderal community could develop. It provided some interesting species such as *Melilotus altissima* ('tall melilot'), a declining species in Dublin and *Hypericum perforatum* (common St John's wort) occurring in its only Dublin city locality. This waste ground has been long established and a number of willow trees are present. As it is outside the region of greatest urban redevelopment it will probably survive some time and be worth visiting in future years.

BUDDLEJA DAVIDII

33

Cultivated plants

Micheline Sheehy Skeffington

One of the interesting elements of an inner city flora is the variety of species which are cultivated, or were so in the past, which in many cases have become fully naturalised in this albeit man-made habitat.

Of the 80 or so species in this flora which can legitimately be called cultivated species, almost half are ornamental garden plants, mostly herbaceous flowers. Some of the most attractive include stock, Californian poppy, evening primrose, mulleins, marigolds and campanulas. The frequency of these may be due to their use as bedding plants. Many are short-lived and are sown annually, particularly in public flower-beds, and are removed as they go to seed. They may seed themselves in that area or become temporarily rooted on waste grounds where they have been discarded. Many are usually not very hardy and cannot compete against vigorous weeds: their survival relies on the occurrence of open ground in their 'wild' habitat and on frequent renewal from gardens.

The bulb and corm-forming species such as bluebells and narcissus are more long-lasting, but do not always flower. One of the commonest is montbretia, which sometimes produces a bright orange flower that is a frequent sight along road-sides in the west of Ireland where it has spread quite rapidly.

Other species which were originally garden escapes are now very common throughout the city. The most conspicuous is buddleja, a butterfly-attracting shrub that was introduced from China during the last century, but has become widely established in Dublin's inner city, apparently relatively recently (since the 1960s). Colgan, in his *Flora of the County Dublin* (1904), does not mention it and it appears not to have been obtrusive enough in the county to merit inclusion in the

D.N.F.C. 1961 *Supplement to Colgan's Flora of the County Dublin*. It is now however extremely abundant throughout the city.

Two other successful inner city plants are the snapdragon (*Antirrhinum*) and the wall valerian (*Centranthus ruber*). Though snapdragons are still planted in the city, the wall valerian is probably rarely deliberately introduced into gardens nowadays. These two species, with buddleja, are successful because they can colonise walls and corners with virtually no soil. They all produce ample small or plumed seeds and from their often lofty vantage-point can spread these some distance on the wind.

Another group of escaped plants includes those cultivated for culinary use. Two families in particular are well represented, the Cruciferae and the Umbelliferae. The first contains many weed species, but also the group *Brassica oleracea*. This name covers a variety of garden vegetables such as cabbage, cauliflower, brussels sprouts, broccoli, etc. Plants frequently regenerate from discarded stalks on waste grounds and may sometimes set seed. Wild turnip (*B. rapa*), another crucifer, is used as a fodder crop and seed is often scattered from lorries and stores. Horse-radish (*Armoracia*) is a particularly vigorous weed once established. It is grown for its root, but is less commonly seen today as it is not often cultivated.

In the Umbelliferae a number of vegetable crops are represented, such as carrot, fennel, celery and parsnip. Carrot and celery are native Irish species but most occurrences in the inner city no doubt arose as a result of discarded cultivated plants. Fennel is more frequently grown as a herb than as a vegetable. It was formerly used medicinally. Alexanders (*Smyrnium olusatrum*) and bishop's weed or goutweed (*Aegopodium podagraria*), two other members of the Umbelliferae, were also formerly cultivated, alexanders as a type of celery that was much used in the 18th century, and goutweed, grown for its medicinal properties during the Middle Ages, as its name might suggest. It was later grown as a pot herb but is now merely a troublesome weed. Thus, though no longer cultivated, both of these plants may have persisted in the

35

Dublin area since the days when they were commonly grown.

Fruit trees and shrubs are found occasionally in waste grounds, though they are regretfully generally non-fruit-bearing. Apple saplings are quite frequently seen, springing up from apple cores. Raspberries can spread considerable distances by means of underground runners and often their seeds are dispersed by birds. Gooseberries also have their seeds dispersed by birds and occasionally arise in a waste ground.

Two other plants abundant in the city are the potato and the tomato, both in the Solanaceae. They are often quite healthy and seem to thrive in the waste ground environment. Renewed regularly from gardens and shops and with house-hold rubbish, they produce tubers or fruit which could feed the less choosy gatherer.

Of the trees naturalised in the city, few survive past the seedling stage. Rarely are seen young saplings of London plane or black poplars, although they are both very common street trees. The Osier (*Salix viminalis*) was planted and pollarded for its long straight shoots which were used in basket making and in thatching cottages. Along the Grand Canal there remain a few specimens that may have been used for these purposes. Colgan (1904) noted that this plant was frequent throughout the county in his time but that it never seemed to have become naturalised.

Some of the grass species occurring in the inner city are included in commercial seed mixes for forage and lawns. Many specimens of the two species of rye-grass (*Lolium perenne* and *L. multiflorum*) found in the city can be seen on roadsides, dispersed by transport vehicles. Wheat, barley and oats also grow occasionally in similar places, and are par-ticularly abundant near the docks and along the roadsides of important transport routes, such as the Grand Canal, no doubt originating from grain fallen from lorries.

Canary-grass, an annual, grows easily from bird seed and its relative abundance shows the frequency with which this seed is scattered. Other species which may also thus have been introduced to the city are cabbage, flax, poppy seed and Indian hemp or *Cannabis*, though cultivation of this last

36

plant in waste grounds is not unknown! Flax is also grown as a decorative plant and may equally occur as a garden escape. Poppy seed is also used in confectionery, generally that of the opium poppy, *Papaver somniferum*. This species has narcotic properties as its names indicate, including its Irish names of 'codlaidín' and 'lus an chodail', but was probably not used medicinally or otherwise in Dublin. Although annual, it can persist for many years in an area as it usually sets good and abundant seed and is quite frequent in the inner city.

There is also a small group of plants which have historical connections with Dublin. The first, *Reseda luteola*, is a native plant known as weld or dyer's rocket. As its names imply, it was used to obtain a yellow dye and was planted in the Dublin district in the eighteenth century for this purpose (Colgan 1904). At that time it was already spreading onto adjacent waste ground. The plants, abundant still in the Liberties, may be derived from this original stock introduced so many years ago.

The three other plants are not native but are frequently found throughout the country, growing adjacent to dwellings. All members of the Compositae, they are aromatic and are popular in herb gardens. Wormwood (*Artemisia absinthium*) has medicinal properties and was also used to dispel insects, especially fleas. Colgan refers to its abundance in 1893-1904 near Lord Edward Street and Nicholas Street. He suggests it spread there from window boxes of now demolished tenement houses. It is still common in this area today, but has not spread to other parts of the city.

The other herbs are still frequent in gardens, the equally aromatic tansy (*Tanacetum vulgare*) and feverfew (*T. parthenium*). Both have medicinal properties, feverfew being described in *An Englishman's Flora* (Grigson 1975) as 'the aspirin of the herbal era', but only tansy appears to have been grown in Dublin in the past. It was 'not infrequent' near dwellings in Colgan's time but nowadays is the rarer of the two in the city. Feverfew has increased in popularity as an ornamental plant, and is now naturalised in many areas of the city, apparently spreading more easily than tansy.

37

Feverfew is one of the more recently naturalised cultivated plants in Dublin's inner city and thrives in this environment. Many of the other cultivated plants only survive in the city through constant reintroduction. It is fortunate that those plants that are successful are often attractive to us and to many insects, including butterflies (*Buddleja*) and moths (white form of wall valerian). Long may they flourish!

Weeds

John R. Akeroyd

Many of the species recorded in Dublin city during the compilation of this Flora have belonged to that group of plants known as 'weeds'. Botanists, agriculturalists, farmers and gardeners have long debated the definition of the word 'weed'. To some, a weed is a common or wayside plant, to others a plant that is a nuisance or grows in the wrong place, i.e. interfering in some way with human activities and industries. The botanist and the weed scientist seek to discover the underlying causes and mechanisms of weed infestations and therefore need to have a more objective definition. However, the concept of nuisance or the injurious nature of weeds is an important component of the scientist's view of weeds.

Weeds are perhaps best seen as *opportunist plants* that are able to move into — and thrive in — the disturbed or seriously modified habitats created by us. They are our 'camp followers', exploiting the habitats in and around our fields and about our dwellings, and they will be with us as long as we till the soil, build and construct and allow the dereliction of unwanted land and buildings. Very broadly, they can be classified into *ruderals* (plants of waste ground and waysides) and *agrestals* (plants of cultivated ground). Once weeds have moved into man-made communities they tend to cause problems. They interfere with the growth of crops and garden plants, they are often untidy and unsightly, they can damage roads (by pushing up tarmac), pavements and buildings; a number of them harbour fungal or bacterial plant disease or are either prickly or poisonous; aquatic weeds block or impede waterways and irrigation ditches. Therefore we have built up an array of techniques to combat weeds which interfere with economic activities, notably agriculture.

In order to persist and to increase in numbers, weeds must

face up to periodic or occasional episodes of catastrophic destruction — by tillage, uprooting, cutting, grazing, burning and spraying with an increasing range of weedkillers. Thus there has been a selection for attributes that allow survival and reproduction, *adaptations* enabling weeds to grow in man-made habitats. By our relentless efforts to exterminate these plants, we have unwittingly promoted their survival. Only the best-adapted variants will grow and reproduce to pass on their features to future generations.

Much has been written about the ways in which weeds exploit their hazardous environment. A number of 'strategies' can be detected:

(i)   A short life history and rapid growth from seed to reproductive phase. A familiar example is the group of common garden weeds that grow up as soon as we cease weeding, especially in wet weather — groundsel, chickweed and shepherd's purse. These plants flower within a few weeks of germination.

(ii) The ability to regenerate from root fragments, often accompanied by brittle roots. Many of the worst weeds have this strategy, including bindweed, couch grass or scutch, creeping thistle and perennial sowthistle. These weeds can grow rapidly from root fragments to flowering.

(iii) The ability to germinate readily when conditions for growth (e.g. heat and light) are favourable, coupled with dormancy and longevity of the seed. The seed of curled dock germinates well in a warm, moist autumn, but seed of this species can survive in the soil for very long periods; one sample showed 8% germination after 70 years' burial! Differences in germination pattern within a species (in response to different environmental factors) have been observed between plants from different habitats, between plants within a population, and between seeds from different parts of the same flowering branch. There is therefore flexible adaptation for immediate and long-term survival.

(iv) Plasticity of the growth and structure of the plant to respond to the demands of the habitat. A plant of common sowthistle growing on top of a wall may only

40

produce three or four heads of flowers: another plant of the same species growing in nearby rich garden soil may produce hundreds of flowers. The most important function of the plant is to reproduce.

(v) Effective dispersal mechanisms, especially adaptation to dispersal of seeds and fruits by wind. Two important groups of weeds in the Dublin area, notably in the inner city, are the thistle, dandelion and groundsel family (Compositae) and the willow-herbs (*Epilobium* spp.). The fruits of many Compositae bear parachutes of spreading, silky hairs; the seeds of *Epilobium* have plumes of long hairs. Weeds that are able to disperse their seeds by wind have the potential to colonise even the smallest and most isolated habitats.

(vi) Flexibility of breeding system, allowing both self- and cross-fertilisation. If a plant is self-pollinated, its progeny will be similar to itself. In a 'difficult environment' such as many weed habitats, this may be of advantage as it will promote the persistence of characters enabling the progeny to survive and will also allow rapid build-up of numbers, perhaps in an environment where there are few insect visitors to carry out pollination and cross-fertilisation. On the other hand it will not be of long-term advantage to a weed to be persistently self-pollinated as this will restrict the genetic diversity of the species, diversity that can be increased by cross-pollination, with the introduction of new characters (some at least adaptive). Weeds tend to show a compromise between retention of existing variation, by self-fertilisation or vegetative reproduction, and the addition of new variation by outbreeding. For example, groundsel and shepherd's purse have small flowers and are rarely visited by insects but are both very variable in form and life history. Evidence suggests that there are occasional episodes of outcrossing which 'bump up' the variation and therefore provide material on which selection can work to produce better-adapted variants.

Thus, as a general rule, weeds are frequently genetically variable species that are well adapted to invade a series of

41

specialised habitats created by human activity. They have evolved or have been recruited from plants that colonise natural open and unstable situations such as strands, river banks, cliffs and animal scrapes. They are usually very demanding of water, light and soil nutrients, especially nitrogen and phosphorus. Some are tolerant of high levels of toxic substances both in soil (metals) and in the atmosphere (e.g. sulphur dioxide).

A major effect of human activity on the weed flora of any region is the often inadvertent introduction of new species, the so-called *aliens* or *adventives*. Many new weed problems arise in this way. A good example of an adventive weed in Dublin is one of the willow-herbs, *Epilobium adenocaulon,* from North America. This has only been observed in Ireland during the last ten years but was found to be abundant throughout Dublin during the summers of 1980 and 1981. Reports from a number of Irish botanists indicate that it is now present in other parts of the county. A high proportion of the inner city flora is adventive. Many adventives are species of open habitats and may eventually become weeds. Others, such as some garden outcasts, fail to persist.

Many weeds are closely related to crops. Many plants that we now see as weeds may once have been cultivated, for example fat hen or lamb's quarters. Other weeds are variants of species that occur on the coasts or in other natural open communities. Curled dock (*Rumex crispus*) is present in Dublin city as the variety *crispus*; on the strands of the County Dublin coast it is a rather dwarfed, compact plant with heavier seeds (var. *littoreus*); on mud of the Shannon estuary and elsewhere it is very tall, with uncrisped leaves and large seeds (var. *uliginosus*). A particularly interesting weed is beet (*Beta vulgaris*). This may be a crop (sugar beet, spinach beet, mangel wurzel, beetroot, chard), a plant of strands (subsp. *maritima*) or a weed, a variant of sugar beet that grows rapidly and flowers in its first year rather than producing a sugar-rich taproot. Many weed variants of common species of seashores, etc. are annuals.

Many flowering plant families contribute weeds, although certain families do have a marked tendency to weediness. In

42

Ireland, a considerable proportion of weeds belong to the cress family (Cruciferae), the dock, sorrel and knotgrass family (Polygonaceae) and to the two major families, the thistle, dandelion and groundsel family (Compositae) and the grasses (Gramineae). Compositae are a major element in the inner city flora, comprising a high proportion of the flora at most sites. The family comprises overall some 15% of the Irish flora.

Many trees, such as sycamore and some shrubs have become weeds. One shrub in particular, buddleja or butterfly bush (*Buddleja davidii*), is a characteristic Dublin shrub which has become a weed. We recorded this species at about every site that we visited. In Britain it has now spread from urban waste ground into the countryside, often in quarries, but in Ireland its distribution is still very localised.

Readers who wish to find out more about weeds should consult the excellent book *Weeds and Aliens* by Salisbury (1961).

Habitats for weeds in Dublin city

There is a wide range of habitats available for plant growth in the inner city of Dublin. The majority of these habitats have been created and maintained by human activity — either consciously by management or unconsciously by neglect — and are thus intrinsically suitable for colonisation by weeds. They fall into two principal classes, relating more or less to the two main groups of weeds, *viz.* ruderals and agrestals, defined above.

(i) Habitats with shallow soils, derived from building materials and a miscellany of inorganic debris. They tend to dry out in summer and are rather nutrient-poor, although often calcareous due to large amounts of mortar. They may be contaminated with toxic substances from discarded waste or aerial pollution.

(ii) Habitats with deeper soils, derived from gardens and frequently enriched by organic rubbish. These soils can contain high levels of nitrogen, phosphorus and other essential nutrients.

Shallow soil habitats

Walls: The vertical faces of walls provide a very specialised habitat, colonised by few plants other than ferns. The tops of walls are richer in species and provide a habitat similar to pavements, car parks and ruined buildings. They are often a long-lasting if inhospitable habitat. A few species may be characteristic: wall speedwell (*Veronica arvensis*), a scarce plant of the inner city, seems to prefer this habitat.

Other habitats with shallow soil: The inner city of Dublin is, regrettably, very much an area of dilapidated houses, broken walls, vacant building sites, car parks, broken pavements and (in the docks) ruined warehouses. These provide a rich mosaic of well-drained, rubble habitats with many niches for the city's weeds. A variety of rubbish, both inorganic and organic, serves to diversify the habitats and may be of a source of introduction of many of the species.

The flora of these often very large decayed areas is derived from pockets of native flora, garden weeds and adventives, from gardens and commerce (e.g. cereals around the docks) and a true 'city element' of plants that tolerate shallow soils and, probably, atmospheric pollution by sulphur dioxide. A classic city plant is the adventive ragwort, *Senecio squalidus,* which is thought to have been introduced to these islands in the 18th century when it was grown in the Botanic Garden at Oxford. It was in Cork city by the earliest years of the 19th century but is a recent arrival in Dublin. In the last ten years it has spread dramatically and can now be found on almost every piece of waste ground in the inner city. Its original home is thought to have been Sicily and it is tempting to suggest that, as well as finding a natural niche on the 'rocks' of Dublin, it can also tolerate the city's sulphur pollution, which is not unlike the volcanic atmosphere of Mt Etna!

Deeper soil habitats

Public and institutional gardens: These larger gardens form the main open spaces of Dublin city. Their flower beds, plant tubs and working areas are a good source for weed records, especially of species that prefer deeper, nutrient-rich soils.

Private gardens: Private gardens carry a similar flora to the larger gardens of the city. In general they are well-weeded but nevertheless provided us with many records. It was remarkable how even the smallest 'pocket handkerchief' of lawn had its quota of daisies.

REYNOUTRIA JAPONICA

46

Mosses and liverworts

Donal Synnott and Peter Wyse Jackson

The most notable bryophyte records for the inner city area are (1) *'Muscus innatus cranio humano* . . . moss growing on a dead man's skull . . . . I took some from skulls upon Custom-House-Quay imported in large Butts from Aghrim' (Threlkeld 1726), and (2) *Hyophila stanfordensis* (Steere) Smith and Whitehouse, on shady soil banks beside paths in St Stephen's Green (Hill 1979). The first is not referable to any modern name; the second was originally described from California and has been found in several places in England in recent years, mainly in the south but extending to Yorkshire. It was found by Dr Peter Pitkin in Dublin in 1978, which remains the only Irish record for the species.

Richardson (1981) has reviewed the literature on the effects of air pollution on bryophytes. The moss flora of an area subject to atmospheric pollution is much impoverished; in general it is sulphur dioxide (SO_2) that causes the most damage to mosses. Gilbert (1968) in England, and Barkman (1969) in the Netherlands, have noted the effects of pollution on the bryophyte flora of urban areas. *Ceratodon purpureus* and *Bryum argenteum* are found to be particularly resistant to SO_2 pollution. Other normally wide-ranging species such as *Tortula muralis*, *Bryum capillare* and *Amblystegium serpens* have a strict requirement for lime-rich soils when subject to pollution.

Our impression of the bryophytes of the inner city is based on visits to the following places: waste ground sites at Montpelier Hill (zone 13), New Street South (zone 6), Charlemont Street (zone 4), Grand Canal Quay (zone 1) and Hanover Street (zone 1); walls and locks of the Grand Canal at Grand Canal Street and Westmoreland Lock (zone 2); shaded clay banks and paths in Iveagh Gardens (zone 4);

mortared walls at New Street, Grand Canal Quay (zone 1) and Iveagh Gardens (zone 4).

On waste ground sites, *Bryum argenteum*, *Ceratodon purpureus*, *Bryum bicolor* and *Barbula convoluta* were usually and predictably present. The relative abundance of *Barbula hornschuchiana* on trampled soil at Hanover Street and Charlemont Street was not expected but it is described in Smith (1978) as 'sometimes locally abundant on waste ground'. *Campylopus introflexus*, an American and Southern Hemisphere species which was first found in Ireland at Howth in 1941 by Miss J.S. Thomson, was found on shaded, humus-rich soil under holly-oaks in Iveagh Gardens and on rotten wood at Montpelier Hill. This species is usually found on dry peat banks or on rotten wood. As with many invading species, *C. introflexus* shows great adaptability to available habitats and has been found also on acid sandy and clay soil, on tree boles (Smith 1978) and on tarmac (Synnott 1982).

The mosses associated with the Grand Canal, *Fontinalis antipyretica*, *Cinclidotus fontinaloides*, *Fissidens crassipes*, *Amblystegium riparium*, *Hygrohypnum luridum* and *Rhynchostegium riparioides*, are all found on limestone boulders on lake shores in the Irish midlands. The limestone blocks which form the banks and locks of the Canals in the city simulate the lake shore conditions in a formal arrangement and allow the lake shore species to enter the urban area.

The following list is based on two short excursions made by the authors. Doubtless there are species to be added to the list and perhaps relevant specimens in herbaria and records in the notebooks of botanists which have been overlooked. However, there are not enough data yet on which to base statements on the state of the moss flora of Dublin in relation to pollution levels.

BRYOPHYTE LIST

This list is intended to indicate what species are found in the city, without systematic reference to zones. However, many are widespread in the city and only the zones where the less common species were found are indicated by their number in brackets after the moss name or site.

Moss nomenclature is according to Smith (1978). Mosses and liverworts generally have no widely-used common names: it was therefore considered that it would not be helpful to include any.

MOSSES

Amblystegium riparium: canal lock just above water level (2).
Amblystegium serpens: shaded clay banks and tree bases (5, 9, 12).
Barbula convoluta: bare soil in open areas.
Barbula cylindrica: gravel path, Iveagh Gardens (4).
Barbula fallax/reflexa: very small plants on mortar under arch of railway bridge, Grand Canal Quay (1).
Barbula hornschuchiana: bare soil in open areas; Hanover Street, Charlemont Street (1, 4).
Barbula recurvirostra: on brick and mortar rubble, Montpelier Hill (13).
Barbula unguiculata: *c.fr., clay and rubble bank, Charlemont Street (4).
Brachythecium rutabulum: grassy patch, Charlemont Street (4), gravel path, Iveagh Gardens (4), grassy patch Westmoreland Lock (2).
Bryum argenteum var. argenteum: waste ground, open sites, common; c.fr., Charlemont Street (4).
Bryum bicolor: on open ground, common, perhaps damper habitats than the last species; bulbils of var. *bicolor* conspicuous at Grand Canal Quay (1).
Bryum cf. ruderale: plants with deep violet rhizoids on clay bank at Iveagh Gardens (4).
Bryum cf. caespiticium: plants with long excurrent nerve, capsules not mature, rubble, Montpelier Hill (13).
Campylopus introflexus: rotten wood, Montpelier Hill (13), humus-rich soil at base of trees, Iveagh Gardens (4).
Ceratodon purpureus: bare soil and rubbish in open places.

---

*c.fr. indicates that fruiting capsules were present on these specimens.

49

Cratoneuron filicinum: gravel path, Iveagh Gardens (4), stony ground, Montpelier Hill (13).

Eurhynchium confertum: on wall rubble, Grand Canal Quay (1).

Eurhynchium praelongum: in *Buddleja* scrub on waste ground, Montpelier Hill (13) and Charlemont Street (4) waste ground among grasses, Grand Canal Quay (1).

Eurhynchium pumilum: clay bank in shade, Iveagh Gardens (4).

Eurhynchium swarzii: clay soil in grassy patch, Charlemont Street (4) base of wall rubble, Montpelier Hill (13).

Fissidens crassipes: on limestone blocks, walls and locks of the Grand Canal, below highest water level (2).

Fissidens taxifolius: clay banks in shade, Iveagh Gardens (4).

Funaria hygrometrica: c.fr., Charlemont Street, on site of fire; c.fr., Montpelier Hill (13).

Hyophila stanfordensis: on shady soil banks beside paths in St Stephen's Green (4) 1978, P. Pitkin.

Hypnum cupressiforme: on limestone wall, Montpelier Hill (13).

Plagiothecium denticulatum: on clay under trees, Iveagh Gardens (4).

Rhynchostegium riparioides: vertical walls of canal lock.

Tortula muralis: on mortared walls and concrete wall caps, common. The species was also found on compacted soil at Hanover Street (1); lower leaves of plants growing in this situation had the nerve excurrent in the normal long hyaline point while on the upper leaves the excurrent nerve was reduced to a short brownish point.

LIVERWORTS

Lophocolea cf. bidentata (L.) Dum.: clay banks in shade, Iveagh Gardens (4).

Lunularia cruciata (L.) Dum. ex Lindb.: gravel path, Iveagh Gardens (4), waste ground, Montpelier Hill (13).

Marchantia polymorpha L.: waste ground, New Street South (6).

Lichens

While the lichen flora was not studied during the current survey, a summary of previously published work is presented here.

A plethora of studies in urban areas points to the toxicity to lichens of sulphur dioxide (SO_2), a principal pollutant from domestic and industrial fuel-burning. In general, both the diversity and the total cover of lichens increase with increased distance from a city centre and the concomitant decrease in air pollution. Studies of this type have been carried out in Ireland by Fenton (1964) in Belfast, Moore (1976) in Dublin and Ní Lamhna *et al.* (1983) in Cork.

Moore (1976) has provided the only account of lichen communities in Dublin's inner city. He investigated the vegetation on Carboniferous limestone blocks and mortar on the walls of 70 churches within a 4 km distance of Christ Church Cathedral and provided abundant data for 21 lichen species of two types: epilithic-crustose (encrusting on stone surface) and foliose (leafy) lichens. Endolithic crusts (those embedded within the substratum) were not included in his study.

While all 21 species were recorded in his outer 2 km wide zones, only eight were found in the inner two and only four of these (*Lecanora conizaeoides, L. dispersa, Candelariella aurella* and *Lepraria incana*) were found within 1 km of Christchurch Cathedral. Other species found within the area covered by the Flora are *Lecidella stigmatea, Scoliciosporum umbrinum, Xanthoria parietina* and *Verrucaria viridula*. No lichens were found on (more acidic) granite walls as this substratum lacks the ability of alkaline rock to 'neutralise' sulphur dioxide.

It may be concluded that the impoverished nature of the

lichen flora in Dublin's inner city is indicative of high levels of air pollution.

Nomenclature follows Hawksworth *et al.* (1980).

Notes on the flora text

Nomenclature

The nomenclature used for all species in the Flora is that of *Flora Europaea*. Synonyms are given in some cases where a previous name is still in common usage. The order of species is also that of *Flora Europaea*.

Common Names

Common names are given for many species that are in common usage in Ireland by amateur botanists. In some cases the choice of name may seem rather arbitrary, but an attempt has been made to avoid the use of the contrived English names used in many wild flower guides.

Distribution

The distribution of each species in the city is given by zone. Where a number is given for a zone it means that the plant has been recorded on at least one occasion in that zone (see map at the front or the back of the Flora).

Irish Names

Irish names have been omitted as they are easily accessible in other works on the Irish flora such as *An Irish Flora* (Webb 1977) and *Census Catalogue of the Flora of Ireland* (Scannell and Synnott 1972).

Abundance

The frequency of each species in the city is given in an ascending scale from very rare to abundant. Broadly speaking the scale can be interpreted in the following terms:

Abundant	) Found in many habitats or adapted to
Very common	) a habitat which is common throughout
Common	) the city.

53

Very frequent	)	The specific habitat is not common or
Frequent	)	widespread, but the plant predictably
Locally frequent	)	occurs whenever this habitat is en-
	)	countered.
	)	The plant occurs in a habitat which is
Rare	)	rare in the city or only a single plant
Very rare	)	or a few plants have been found in
	)	the city.

Notes on each species

Preferred habitat is given in most cases. In some cases, when information is not readily available in other common texts, one or two easily described characteristics are included to separate two closely related species.

Irish status

The status of each species is given for Ireland, whether they are native in Ireland, possibly introduced, probably introduced or certainly introduced. This refers to their status in Ireland as a whole and *not* to their status in Dublin. The authors have been guided in this matter by *An Irish Flora* (Webb 1977).

Plant records

The year of first discovery during the project is given for rare or interesting plants, as well as the initials of the finders. If there were more than a single record for the plant in the city then the zone number is given after the details of locality.

Recorders

In the text initials of recorders are given as below and represent the following people:

JRA	John Akeroyd	PWJ	Peter Wyse Jackson
PHC	Peter H. Carvill	EL	Eimear Lowe
DD	Declan Doogue	JP	John Parnell
KD	Katherine Duff	PR	Paddy Reilly
MH	Mary Hennessy	MS	Maura Scannell
DH	Des Higgins	JS	Jonathan Shackleton
HH	Howard Hudson	MSS	Micheline Sheehy Skeffington

54

Where the initials DNFC have been used it indicates that the record was made on an outing by the Dublin Naturalists' Field Club. The letters **DBN** indicate that the record is from a specimen in the National Herbarium, Glasnevin.

The flora list

PTERIDOPHYTA
EQUISETACEAE

Equisetum arvense x fluviatile (E. x litorale)   Horsetail
$— — — — — — — — — — — — 14$
Native, rare.
This is a common hybrid in Ireland. It frequently occurs
in the absence of one or both of its parents. Royal Canal
banks, '81, DD, PWJ.

E. palustre   Marsh Horsetail
$— — — — — — — — — — — 12 — —$
Native, rare.
Abandoned, marshy railway line at Cabra, '81, DNFC.

E. arvense   Horsetail
$— — — — — — — — — — — 12 — 14$
Native, locally frequent.
Royal Canal banks and railway embankments. The
commonest horsetail in the inner city, it is often in a
drier habitat than the other species.

PTERIDACEAE
Pteris cretica
$— — — — — — — — — 10 — — — —$
Introduced, very rare.
The plant is often cultivated as a house plant in Ireland.
It is native to southern Europe and Asia. Only one record
for this plant in the city, under a grille in Parnell Street,
'81, PR.

56

Pteridium aquilinum   Bracken
1 — 3 4 5 6 7 8 9 10 11 — — —
Native, common.
Common in most parts of the city, often on walls, the
edges of pavements and in basements.

ASPLENIACEAE
Asplenium trichomanes   Maiden Hair Spleenwort
1 — — — — — 7 — 9 10 11 12 13 14
Native, locally frequent.
Occurring in wall crevices but more rarely than the next
species. It is perhaps less drought resistant than others.

A. ruta-muraria   Wall-rue
1 2 3 4 5 6 7 8 9 10 11 12 13 14
Native, common.
Grows very commonly in the crevices of old limestone or
mortared walls.

Phyllitis scolopendrium   Hart's Tongue Fern
1 2 3 4 5 6 7 8 9 10 11 12 13 14
Native, common.
This fern is perhaps the commonest wall fern to be found
in the city. It occurs on many walls especially wherever
there is a leaky drainpipe to give it additional moisture.

ASPIDIACEAE
Dryopteris filix-mas   Male Fern
1 — 3 4 5 6 7 8 9 10 11 — — 14
Native, frequent.
Common on walls and in the basements of houses throughout the city. It is often abundant on damp old walls,
where water flows from broken drainpipes. Many specimens are young and undeveloped and therefore difficult
to identify.

D. pseudomas (*D. borreri*)

— — — — — — 8 — — — — — —

Native, very rare.

It was not described as a distinct species until recently and so was overlooked by Colgan. It is however much rarer in the Dublin region than the previous species. In one locality near the Royal Canal at Sheriff Street, '80, DNFC.

Polypodium vulgare agg.   Common Polypody

1 2 — — — — 7 — — — 11 — 13 14

Native, frequent.

Frequently encountered on old walls.

The three species found in Ireland can only be distinguished with certainty on miscroscopic examination of the sporangia. The records could not thus be checked in this survey.

DIOCOTYLEDONES
SALICACEAE

Salix atrocinerea   Sally

— — 3 — — — 7 8 — — — — — —

Native, occasional.

This plant is found occasionally in long-established waste grounds throughout the city. The mode of dispersal in Dublin is not known, but most *Salix* species root freely from cuttings. Waste ground in Earlsfort Terrace, '81, (3), PWJ, JRA.

S. aurita   Eared Willow

— — — — — — — 8 — — — — — —

Native, rare.

In one long-established waste ground near the Royal Canal at Sheriff Street, '80, DNFC.

S. caprea   Goat Willow

1 2 3 — — — — 8 — — — — — 14

Native, occasional.

Occasional in long-established waste grounds and on the

banks of the two canals. This is a fine willow when mature and may have been planted in some places.

S. viminalis   Osier
— 2 — — — — — — — — — — — —

Probably introduced, rare.
This species was much used for basket-making and may have been planted for this purpose by the canal. It flourishes on damp ground. Grand Canal bank, near Dolphin's Barn Bridge, '81, PWJ, MSS.

Populus alba   White Poplar
— — 3 — — — — — — — — — — —

Introduced, rare.
This tree is occasionally planted in the city and suckers freely from the base. Earlsfort Terrace waste ground, '81, PWJ.

P. nigra   Black Poplar
— 2 — — — — — — — — — — — —

Introduced, rare.
The commonest variety found is var. *italica*, the Lombardy Poplar, a tall, narrow, quick growing tree. Some varieties are very tolerant of atmospheric pollution and are therefore very suited to the city environment. It produces suckers freely from its base which may persist after the parent plant has been removed. Grand Canal banks, '81, DNFC.

P. x canadensis (*P. nigra x deltoides*)   Hybrid Black Poplar
This hybrid black poplar is a commonly planted street tree and is often confused with the previous species. It does not seed itself in the city. Fertile seeds on any poplar species are rare in Dublin.

BETULACEAE
Betula pendula (*B. verrucosa*)   Birch
— — 3 4 — — — — — — — — —
Native, rare.

Commonly planted. Seedlings are occasionally found near planted trees. It does not appear to naturalise itself in wastegrounds. Certainly planted in zones 2, 3, 4, 5, 6 and 8. Seedlings under trees in Trinity College, (4), '81, MSS.

B. pubescens Birch
— — 3 4 — — — — — — — — — —
Native, rare.
Occasionally planted but rarely naturalised in waste grounds. Hybrids between this and the previous species are common in Ireland but apparently not in Dublin.

Alnus glutinosa Alder
— — — 4 — — 7 — — — — — — 14
Native, rare.
Unlikely to have been planted in zones 7 and 14. The station in zone 4 is more suspicious as it is the site of an old garden. This tree grows in hedgerows and marshy places. Waste ground off Charlemont Street, (4) '81, PWJ. Liffey Walls (7), '82, MSS.

CORYLACEAE
Corylus avellana Hazel
— — 3 — 5 — — — 9 — — — — —
Native, rare.
Rather rarely planted in the city. May occasionally grow from hazel nuts.

FAGACEAE
Fagus sylvatica Beech
1 — — — — — — — — — — — — —
Introduced, rare.
Commonly planted in all zones. Very rarely self-seeded. Seedling near Grand Canal dock, '80, PWJ.

Quercus cerris Turkey Oak
— — 3 — — — — — — — — — — —
Introduced, rare.

60

Occasionally planted but very rarely self-seeded. Seedling found at Earlsfort Terrace, '79, DH, PWJ.

ULMACEAE

Ulmus glabra   Wych-Elm
1 2 — 4 — — — — 9 — — — 13 —
Native, occasional.
Generally found in long-established waste grounds.

U. procera   English Elm
— 2 — 4 5 — — — — 10 11 12 — —
Introduced, occasional.
This species is always planted. The tree suckers freely and specimens originating as suckers may persist after the parent has died, giving the impression that it is self-sown. Occasionally found in long-established waste grounds.
The planted population of elm in the city has been devastated in recent years by the rapid spread of Dutch Elm Disease. As a result few elms are now planted. In a number of places on the banks of the Grand Canal, in water, very strikingly coloured pink roots can be seen, probably from one of the two species of elm to be found nearby.

MORACEAE

Ficus carica   Fig
— — — — — — — — — — — — — 14
Introduced, very rare.
Grown in many old gardens in Ireland but very rarely naturalised. This unexpected plant was found in a single station in the inner city where it had obviously seeded itself or regenerated from a discarded plant. The site has since been cleared and the small tree is gone now. Near Mountjoy Prison, on the bank of the Royal Canal, '80, HH.

URTICACEAE
Urtica dioica   Nettle
   1 2 3 4 5 6 7 8 9 10 11 12 13 14
   Native, abundant.
Commonly found in most waste grounds, neglected
gardens and on the canal banks. It grows particularly
well on phosphate-rich soils. An old denizen of Dublin,
seeds have been found in the deposits at Wood Quay.
A non-stinging variant has been found close to Cross
Guns Bridge (14), '80, HH.

U. urens   Annual Nettle
   1 2 − 4 − 6 − − − 10 11 12 − −
   Native, occasional.
Found on recently turned or cultivated soils. A smaller
species than its perennial relative, *U. dioica*.

Parietaria diffusa   Wall Pellitory
   1 − − − 5 6 − 8 − − − − 13 −
   Native, locally frequent.
Not as common in Dublin as it is in other Irish cities.
On walls close to the sea or the Liffey River only.

Soleirolia soleirolii (*Helxine soleirolii*) Mother of Thousands
   1 − 3 4 5 − − − 9 − − − 13 −
   Introduced, occasional.
This garden escape occurs frequently close to old gardens
where it is grown as a ground cover plant. It occurs
both on open soil and on walls. Native to the islands
of the Western Mediterranean.

POLYGONACEAE

Polygonum aviculare   Knotgrass
   1 2 3 4 5 6 7 8 9 10 11 12 13 14
   Native, abundant.
Found commonly in most city habitats but it prefers
open plant communities. (Figure p. 65).

62

P. arenastrum
1 2 3 4 5 6 7 8 9 10 11 12 13 14
Native, very common.
Grows in similar habitats to the last species, although it is most characteristic of dry, trampled communities, e.g. paths and playing fields. (Figure p. 65)

P. persicaria   Spotted Persicaria
1 2 – 4 5 6 7 8 9 10 11 12 13 14
Native, frequent.
Occurs on recently disturbed and open habitats on damp soils. It is widespread in the city. (Figure p. 65)

P. lapathifolium
1 – – 4 5 – 7 8 – – – 12 13 14
Possibly introduced, occasional.
Frequently confused with the previous species. A number of specimens collected in the city have been difficult to assign to this or the other species. It grows mainly on disturbed soils. Colgan notes that this species has a preference for peaty soils. It appears to be less widespread than in Colgan's time, (DNFC Suppl. 1961).

P. amphibium
1 2 – – – – 7 – – – – – – 14
Native, occasional.
This species requires a damper soil than is usual in urban waste grounds. Many of the plants seen had black markings on the leaves, like those of *P. persicaria*, a characteristic rarely mentioned in floras. Found only on the canal banks and in the filled-in Grand Canal Harbour (7).

P. amplexicaule
– – 3 – – – – – – – – – – –
Introduced, very rare.
One record of this plant as a casual weed only. It is naturalised in western Ireland but is much rarer elsewhere. Fitzwilliam Square, '80, KD.

1. Polygonum aviculare
2. P. arenastrum
3. P. persicaria
4. Chenopodium album
5. Fallopia convolvulus
6. Atriplex patula

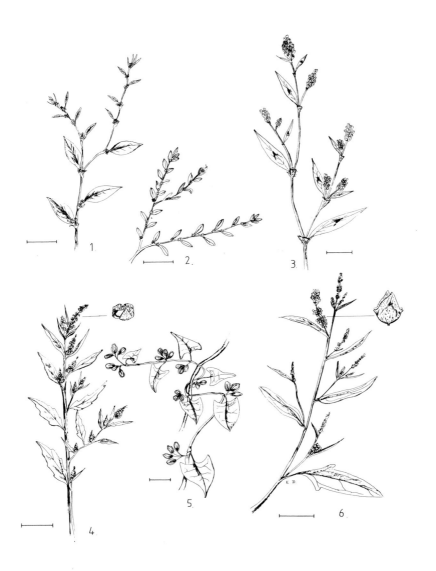

Bilderdykia baldschuanica (*Polygonum baldschuanicum*)
Russian Vine
– – – – – – – – 9 – – – – –
Introduced, rare.
Frequently found as a wall plant of gardens in the suburbs of the city. It is rarely if ever naturalised but frequently grows rampant. In zone 9 it occurs in an open waste ground where it is unlikely to have been planted but is possibly only a discard from a garden.

Fallopia convolvulus (*Bilderdykia convolvulus, Polygonum convolvulus*)
Black Bindweed
1 – – 4 5 6 7 – 9 – – – – –
Possibly introduced, frequent.
Frequent in the region close to and south of the Liffey in waste grounds, rarer or absent elsewhere. A plant of tilled ground rather than of waste grounds. It may have been more common around Dublin in Colgan's time. (Figure p. 65)

Reynoutria japonica (*Polygonum cuspidatum*)
Japanese Knotweed
1 2 3 4 5 6 7 8 9 10 11 12 13 14
Introduced, common.
A garden escape, now established in many waste grounds. Difficult to eradicate once established. One of the most characteristic plants of ruderal sites in Dublin city. A vigorous plant growing up to 2 m tall on some occasions. (Figure p. 46)

Rumex acetosella
Sheep's Sorrel
1 – – – 5 – – – – 10 – – – 14
Native, rare.
Common in Ireland on peaty soils.

R. acetosa
Common Sorrel
– – – – – 6 – – – 10 – – – –
Native, rare.
Waste grounds and bases of walls and pavements. This species is usually found in moist meadows and shady

places, a habitat infrequent in the inner city.

R. crispus   Curled Dock
1 2 3 4 5 6 7 8 9 10 11 12 13 14
Native, very common.
Present in many waste grounds and ruderal sites. Also
on the canal banks. All material in Dublin city is var.
crispus, a common weed variant. (Figure p. 69)

R. conglomeratus   Clustered Dock
1 2 — — 5 6 7 — — 10 — 12 13 14
Native, frequent.
Frequent in waste grounds throughout the city. (Figure p. 69)

R. sanguineus   Wood Dock
1 2 3 4 5 6 7 8 — 10 — 12 — 14
Native, frequent.
Waste grounds. In similar places to the previous species
and often with it. (Figure p. 69)

R. obtusifolius   Broad-leaved Dock
1 2 3 4 5 6 7 8 9 10 11 12 13 14
Native, abundant.
This is the commonest species of *Rumex* found in the
city, occurring in most habitats. Colgan also notes that
it is the commonest waste land dock.

CHENOPODIACEAE
Beta vulgaris   Beet
1 — — — — — — — — — — — — —
Introduced, rare.
Possibly spinach. A cultivated crop escape. In a waste
ground near St Andrew's School, Pearse Street, with
other garden discards, '81, PWJ, JRA.

Chenopodium album   Fat Hen, Goosefoot, Lamb's Quarters
1 2 — 4 5 6 7 8 9 10 11 12 13 14
Introduced, common.
It is found in many waste grounds throughout the city,

1. Rumex crispus
2. R. sanguineus
3. R. conglomeratus
4. Euphorbia helioscopia

especially on recently disturbed ground. A very variable species that is one of the first colonisers of a new waste ground. (Figure p. 65)

Atriplex patula   Orache
1 — 3 4 — 6 7 — 9 10 11 — 13 14
Possibly introduced, frequent.
Frequently found on open ground and waste grounds in similar places to the previous species. (Figure p. 65)

A. hastata
1 — — 4 5 — 7 8 — — 11 — 13 14
Native, frequent.
Frequent on open ground and recently disturbed ground, in waste places. Less common than the previous species. It occurs often accompanied by the previous two species, all three being characteristic of nutrient-rich soils.

CARYOPHYLLACEAE
Arenaria serpyllifolia
1 — — — — — — — — 10 — 12 — —
Native, rare.
Grows mainly on bare sandy ground and old limestone walls. It occurs mostly on the ground in Dublin.

Stellaria media   Chickweed
1 2 3 4 5 6 7 8 9 10 11 12 13 14
Native, abundant.
One of the commonest city weeds in practically every habitat. It is a very successful coloniser of bare or cultivated ground and may indicate nutrient enrichment. (Figure p. 91)

S. holostea   Greater Stitchwort
— — — — — — — — — — 11 — — —
Native, very rare.
A hedgerow species found in the grounds of St Brendan's Hospital, '81, DD.

70

Cerastium fontanum
1 — 3 4 5 6 7 8 — 10 11 12 — —
Native, locally frequent.
A weed of waste grounds and pavements in the city.
(Figure p. 91)

C. glomeratum
— — — — — — 7 — — — — 12 13 —
Native, rare.
A weed of cultivated and waste grounds.

Sagina procumbens   Procumbent Pearlwort
1 2 3 4 5 6 7 8 9 10 11 12 — —
Native, very common.
This is the commonest Pearlwort species. It is one of the
few plants which is abundant in the cracks of pavements
and steps and on walls.

S. apetala   Annual Pearlwort
1 2 — — — — 7 8 — — 11 12 — 14
Native, occasional.
This species is easily confused with the last one. They can
look very similar especially on trampled ground. It is
however much less common than the last.

Spergula arvensis   Corn Spurrey
— 2 — 4 — — — — — — — — — —
Introduced, very rare.
A weed of cultivated ground, rare in Dublin city and
preferring a sandy acid soil. Allotments on the banks of
the Grand Canal near Dolphin's Barn Bridge, (2), '81,
MSS, PWJ. Irish Times Car Park, Hawkin's Street, (4).
'79, PHC, PWJ.

Silene vulgaris subsp. vulgaris   Bladder Campion
— — — — — — — 8 — — — 12 13 —
Native, rare.
Absent from most of the inner city except the north-
west part, but also in a waste ground close to the Royal
Canal at Sheriff Street (8), '80, DNFC. Roadside at
Fassaugh Road (12), '80, PWJ.

71

S. alba   White Campion
$- 2 - - - - - - - - - - - 13 -$
Native, rare.
On the canal banks, Grand Canal Street (2). '79 MSS,
PWJ. Though noted as a casual inland by Colgan, he has a
record for this species near the fifth lock of the Grand
Canal. Also in a waste ground at Montpelier Gardens
(13), '80, DNFC.

Saponaria officinalis   Soapwort
$- - - - - - - - - - - - - 14$
Introduced, very rare.
Found in one station only in the inner city where it
occurs as the double-flowered form, variant 'flore pleno'.
No doubt a garden escape. Near Cross Guns Bridge,
'80, HH.

NYMPHAEACEAE
Nuphar lutea   Yellow Water Lily
$- 2 - - - - - - - - - - - 14$
Native, frequent in the canals.
Occurs only in the two canals in Dublin city.

CERATOPHYLLACEAE
Ceratophyllum demersum   Hornwort
$- 2 - - - - - - - - - - - -$
Only in the Grand Canal where it is very frequent and in
some places abundant. A recent introduction to the canal.
Its spread may have caused the decline of other aquatic
plants in the canal since its introduction.

RANUNCULACEAE
Clematis vitalba   Old Man's Beard, Traveller's Joy
$- - 3 - - - - - - - - 12 - -$
Introduced, rare.
A garden plant that was recorded in two zones only,
becoming sometimes rampant, but is probably always
planted and never naturalised.

Ranunculus repens   Creeping Buttercup
  1 2 3 4 5 6 7 8 9 10 11 12 13 14
  Native, abundant.
  A common plant of most city habitats, especially grassy
  places and closed plant communities. It spreads rapidly
  by means of stolons and is a more characteristic plant of
  waste grounds than the next species.

R. acris   Meadow Buttercup
  − 2 3 4 5 − 7 8 − − − 12 13 14
  Introduced, occasional.
  A plant of long-established waste grounds, grassy places
  and banks.

R. bulbosus   Bulbous Buttercup
  − 2 − − − − − − − − − 12 13 −
  Native, rare.
  Found in similar types of habitat to the last species, but
  is much less frequent preferring a sandy substrate.

R. sceleratus
  − 2 − − − − − − − − − 12 − 14
  Native, rare.
  Grows beside the two canals and in a wet abandoned railway line by Cabra Road, (12), '80, PR; (2), '80, MSS
  (14), '80, HH.

R. ficaria   Lesser Celandine
  − − − 4 − − − − − − − − −
  Native, very rare.
  It is a common garden weed of shady habitats frequently
  as the subspecies *bulbifera*, which has bulbils present in
  the axils of the leaves. Found in a single station only in
  the inner city, in an old garden off Charlemont Place,
  beside the Grand Canal, '81, PWJ.

PAPAVERACEAE

Papaver somniferum   Opium Poppy
  1 − − − 5 6 7 8 − 10 11 − 13 14
  Introduced, occasional.
  This plant is common in waste grounds. Grown in some
  gardens for ornamental purposes from which it spreads
  easily by seed. Little grown in Dublin for medicinal or
  narcotic purposes.

P. dubium   Poppy
  1 − − 4 − 6 − − − − − − 13 −
  Probably introduced, rare.
  Occasionally found in waste grounds throughout the city,
  but never abundantly. It prefers a recently disturbed
  habitat.

P. lecoquii
  − − − − − 6 − − − − − − −
  Introduced, very rare.
  This species is very similar to the previous species but can
  be readily distinguished by its orange coloured latex.
  Possibly overlooked in some instances in the city, but
  certainly never frequent. Waste ground off Clanbrassil
  Street Lower, '81, DNFC.

P. rhoeas   Corn Poppy
  1 2 3 4 5 6 7 − − 10 11 − − 14
  Possibly introduced, frequent.
  This is the commonest poppy found in the city. It is
  frequent on many wastegrounds but prefers cultivated
  or disturbed soils. It is easily distinguished from *P.
  dubium* by the out-standing hairs on the stem. *P. dubium*
  has flattened hairs.

Meconopsis cambrica   Welsh Poppy
  − − 3 − − − − − − − − − −
  Native, rare.
  Although native in Ireland it is certainly introduced to

74

Dublin, where it must be a survivor from an old garden. Earlsfort Terrace waste ground, '81, PWJ, JRA.

Eschscholzia californica   Californian Poppy
    — — — — — 6 — — — — — — — — —
Introduced, very rare.
A garden escape and never more than a casual. Found on one waste ground only in the city. Off Clanbrassil Street Lower, '81, JP.

Fumaria officinalis   Fumitory
    1 2 — 4 5 6 7 — — — 11 — 13 14
Possibly introduced, frequent.
A common species of waste places and cultivated ground especially near the east coast.

F. muralis
    1 2 — 4 — — — — — — 11 — — —
Possibly introduced, occasional.
Occurs in similar habitats to the previous species.
Fumaria species are difficult to identify because they are morphologically very variable and there are few good diagnostic characters to separate them. Both these species of *Fumaria* occur as weeds of recently disturbed ground, often in flower-beds and recently laid lawns.

CRUCIFERAE

Sisymbrium orientale
    1 2 3 4 5 6 7 8 — 10 11 — 13 —
Introduced, common.
A common weed of waste grounds, often with a large population size and growing with other crucifers. The DNFC Supplement notes the rapid spread of this species since Colgan's time and it appears to have become widely distributed throughout the city since 1961. (Figure p. 79)

S. officinale   Hedge Mustard
    1 2 3 4 5 6 7 8 9 10 11 12 13 14
    Native, abundant.
    A very common weed of waste grounds and other ruderal
    habitats. This scruffy plant with small yellow flowers is
    one of the most frequently encountered plants in the city.
    S. irio which was recorded in Dublin in Colgan's time is
    apparently absent in the inner city and is probably extinct
    in Dublin. (Figure p. 79)

Arabidopsis thaliana   Thale Cress
    — — — 4 — — — — — — — 12 — —
    Native, rare.
    This weed of disturbed ground and flower-beds is ap-
    parently rare in Dublin city, but possibly overlooked in
    a few cases due to its early flowering and inconspicuous
    appearance. It may sometimes be introduced with bedding
    plants or shrubs as it is a common weed of nurseries and
    garden centres in the Dublin area. Waste ground off
    Fassaugh Road, (12), '81, PR. Outside Trinity College
    gymnasium, (4), '79, PWJ, MSS.

Hesperis matronalis   Dame's Violet
    — — — — — — — — 9 — — — — —
    Introduced, very rare.
    A common country cottage-garden escape, this plant was
    never common in Dublin.

Cheiranthus cheiri   Wall-flower
    — — — — — — — — 9 — — — — —
    Introduced, very rare.
    A single station only, on an old wall of a house on
    Berkley Road. A garden escape that occasionally becomes
    well established in the wild, '79, **DBN.**

Matthiola incana   Stock
    — — — 4 — — — — — — — — — —
    Introduced, very rare.
    This common garden plant is apparently rarely self-sown

and hardly ever encountered outside the garden environment. Clonmel Street, beside the pavement, '80, PWJ.

Barbarea vulgaris   Yellow Rocket, Winter Cress
1 — — — — — — — — — 11 — — —
Native, rare.
Although this species is frequent throughout the country in waste places and on roadsides, it is not common in Dublin city. Sir John Rogerson's Quay, (1), '81, DD, JRA, PWJ. Grangegorman, in a waste ground, (11), '81, DD, JRA.

Armoracia rusticana   Horse-radish
— — — — — 6 — — — — — — —
Introduced, rare.
A common discard from gardens during the last century. It is less cultivated today and therefore less frequently encountered in the wild. Waste ground near Rialto Bridge (6), '80, DNFC. Royal Canal bank (14), '80, HH.

Nasturtium microphyllum   Water-Cress
— — — — — — — — — — — — — 14
Native, rare.
Grows in wet habitats by the Royal Canal bank, but apparently not in or near the Grand Canal.

N. officinale   Water-Cress
— 2 — — — — — — — — — — — 14
Native, rare.
Present in both the Royal and Grand Canals in similar habitats to the last species.
The two species of *Nasturtium* can be distinguished by means of seed capsule characters. In *N. officinale* the siliqua is 13-18 mm long and the seeds are in two rows, whereas in *N. microphyllum* the siliqua is 16-22 mm long and the seeds are in a single row in each loculus. Hybrids between the two species are rather frequent and have been recorded in the Royal Canal, (14), '81, DD. The hybrid spreads vegetatively.

1. Sisymbrium orientale
2. S. officinale
3. Brassica oleracea
4. Reseda luteola
5. Brassica rapa

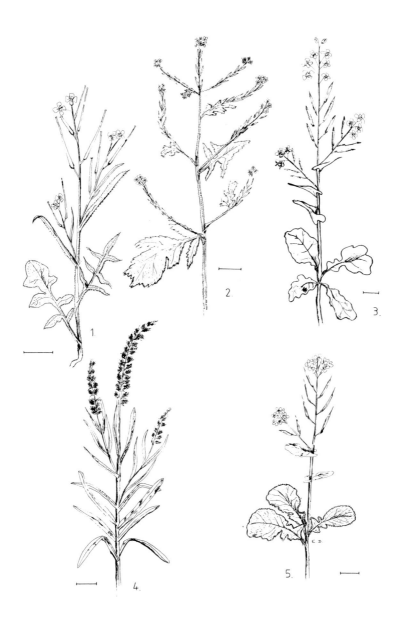

Cardamine pratensis   Lady's Smock

— 2 — — — — — — — — — — — — —

Native, rare.

A plant of marshy ground and wet meadows, it is rare in
Dublin even on the banks of the Grand Canal.

C. hirsuta   Hairy Bitter-Cress

1 2 3 — — — 7 — — — — — — —

Native, occasional.

A weed of waste places and gardens that is nowhere
common in the city. Found on the railway track at Westland Row Station with *Cochlearia danica*, '80, PWJ.

Cochlearia danica   Danish Scurvy-grass

1 — — — — — — — — — — — — —

Native, very rare.

Although this plant is primarily a coastal plant of shingle
beaches and rocks and walls by the sea, in a number of
places in Ireland it grows inland on railway lines and walls.
The record at Westland Row is not therefore entirely unexpected as this plant is quite frequent all along the coast,
north and south of Dublin, often close to, or on, the
railway line, '80, Tom Curtis.

Capsella bursa-pastoris   Shepherd's Purse

1 2 3 4 5 6 7 8 9 10 11 12 13 14

Native, abundant.

An abundant weed of waste grounds, especially recently
disturbed ones, pavements, flower-beds and most other
ruderal habitats. It is an extremely variable plant in Dublin
as elsewhere, with much variation in morphology, size
and life history, and has been the subject of much study.

Lepidium latifolium

— — — — — — — — — — — — — 14

Probably introduced, very rare.

This species is quite frequent on the coast north of the
inner city. The banks of the Royal Canal near its junction
with the Liffey is its only station in the inner city. Several

80

other species of *Lepidium* have been recorded in County Dublin as casuals in the past, '81, DD, PWJ, JRA.

Coronopus squamatus   Swine-cress
1 − − − 5 − 7 − 9 − − − − 14
Possibly introduced, occasional.
Found usually in waste grounds and trampled places. It is the rarer of the two species of *Coronopus* in the city, but was the commoner in former times.

C. didymus   Lesser Swine-cress
1 2 3 4 5 6 7 − − − 11 − 13 14
Introduced, common.
Grows in waste places and on trampled ground, in similar places to the last species but more commonly. It is more frequent on the south side of the Liffey than on the north. It appears to have spread widely since Colgan's time at the turn of the century, when he recorded it as rare.

Diplotaxis muralis   Wall Rocket, Stinkweed
− − 3 − − − − − 9 − − − 13 −
Introduced, rare.
This crucifer is occasional in parts of Dublin, often occurring by railway lines. Heuston Station is the major centre for the plant but it does not seem to have invaded the inner city to any great extent. Building site off Baggot Street, (3), '80, PWJ. Waste ground near the Mater Hospital, (9), '81, DNFC.

Brassica oleracea   Wild Cabbage
1 − 3 4 − 6 7 8 9 10 − − 13 −
Introduced, frequent.
This species has many cultivated races, such as cabbage, brussels sprouts, cauliflower, etc. These are commonly discarded with household rubbish and often become established in waste grounds throughout the city. All plants of this species found in Ireland are introduced in this way or by horticultural seed. (Figure p. 79)

B. rapa   Wild Turnip

1 2 3 4 5 6 7 8 – – 11 12 13 14

Possibly introduced, frequent.

This conspicuous waste ground weed is quite frequent
throughout the city and is easily spotted due to its early
flowering. It is one of the first colonisers of waste grounds
and, though an annual, persists from year to year in the
same place, whenever the plant community remains open.
(Figure p. 79)

B. nigra

– – – – 5 6 7 – – – – – – –

Native, rare in most parts of the city.

This plant occurs in waste grounds and often attains a
large population size, persisting from year to year in the
same site. Generally found accompanied by *Rapistrum
rugosum*. It was becoming rare in Colgan's time in 1901.
In the DNFC Supplement of 1961, it is said to be almost
extinct. It seems that a local source of seed is maintaining
populations in the inner city. Found in only one small
region of the city with its centre around St Patrick's
Cathedral.

Sinapis arvensis   Charlock

– – – 4 – 6 – – – – – – 13 14

Probably introduced, occasional.

Found occasionally in waste grounds and on roadsides
but never as more than a few individuals in each case.
This plant has not really become established to any
extent in Dublin city. A plant of cultivated fields, it was
sold as 'corn-cail' in Dublin in the early eighteenth century.

Rapistrum rugosum

1 – – – 5 – 7 – – – – – 13 –

Introduced, locally frequent.

A waste ground weed often growing with *Brassica rapa*.
It seems to have increased in recent years in the city and
is now firmly established. It persists from year to year in
most of its sites. Its centre of occurrence is around St
Patrick's Cathedral, with one station on the north bank

of the Liffey, (13), '81, PWJ. Sir John Rogerson's Quay (1), '81, PWJ. A large population was found near Christ Church Cathedral (5), '80, PWJ, MSS.

Raphanus raphanistrum   Wild Radish
– – – 4 – – 7 – – 10 – – – –
Introduced, rare.
A weed of cultivated crops that has never really spread within the city. Occurs on a few waste grounds in the city.

RESEDACEAE

Reseda luteola   Yellow-weed, Dyer's Rocket
1 – – – 5 6 7 – 9 10 11 12 – 14
Native, frequent.
A weed of waste grounds, especially in sandy places. It is more common in the western half of the city than in the east. A calcicole, it is also common on limey walls and rubble. It was formerly grown as a source of yellow dye and seems to have been a common weed as early as 1726 (Threlkeld). (Figure p. 79)

CRASSULACEAE

Sedum acre   Wall Pepper
– 2 – – – 6 – – – – 11 12 – –
Native, occasional.
Found on old dry walls infrequently throughout the city. The leaves have a peppery taste when eaten. Surprisingly the garden stonecrop, *S. album*, has not been recorded, even though the DNFC Supplement notes that it was spreading rapidly in 1961.

SAXIFRAGACEAE

Saxifraga tridactylites   Rue-leaved Saxifrage
– – – – – – – – – – – 12 – –
Native, very rare.
Grows on walls and dry sandy places. A calcicole, it is

83

surprisingly rare in the city. It was known from the city in the eighteenth century, being recorded near Kevin Street. Railway line off Fassaugh Rd., '81, PR.

GROSSULARIACEAE

Ribes sanguineum   Flowering Currant

— — — — — — — — — — — — 14

Introduced, rare.
This garden plant, native to North America, is commonly planted in Dublin gardens. The only non-garden record for the plant is on the bank of the Royal Canal, where it may have sown itself, '80, HH.

R. uva-crispa   Gooseberry

— — — 4 — — — — — — — — — —

Introduced, very rare.
A garden plant that is cultivated for its fruit in gardens. The only record for the plant in the wild is in a waste ground off Charlemont Street. '80, PWJ. It may have arisen from seeds dispersed by birds or from a discarded specimen.

PLATANACEAE

Platanus x hybrida   London Plane

1 — — — — — — — — — — — — —

Introduced, very rare.
Although this is a very commonly planted street tree in most of the city zones, it is rarely self-sown even though its seeds are fertile. The origins of this plant are obscure. It is either a cultivar of *P. orientalis* or a hybrid between that species and *P. occidentalis*. It is particularly suited to the city environment as it can tolerate atmospheric pollution well.

Filipendula ulmaria   Meadow-sweet
— 2 — — — — 7 — — — — — — 14
Native, rare.
Found in wet or damp habitats beside the canals. It is rare because of the lack of suitable habitats. Grand Canal Harbour (7), '81, DNFC.

Rubus idaeus   Raspberry
— — — — — — — 8 — — 11 — 13 —
Introduced, occasional.
This is the common cultivated raspberry. It is occasionally discarded from gardens and becomes established in waste grounds, or else it persists in old gardens long after the garden has ceased to exist. It suckers very easily and spreads in this way over a considerable distance.

R. fruticosus agg.   Bramble
1 2 3 4 5 6 7 8 9 10 11 12 13 14
Native, very common.
Established in many waste grounds and derelict gardens. Commonest in sites that have remained undisturbed for a number of years.
This species aggregate is notoriously difficult taxonomically, due to the apomictic nature of its reproduction (seeding without sexual fusion). This means that each plant is genetically isolated and any mutation can give rise to a new variant. Only easily recognisable species have been distinguished in this work.

R. ulmifolius
— — — — — 6 7 — — 10 — — — 14
Native, frequent.
Frequent on limestone soils, rarer on acid ones. It can be distinguished easily by its small leaflets that are white underneath and covered with a downy felt. It is one of the few known true sexual species of bramble. Royal Canal Bank (14), '81, PWJ, DD, JRA. Lane off Dolphin's Barn Road (6), '81, PWJ, MSS.

R. caesius

— — — — — — — — — — — — — 14

Native, rare.

It grows on sandy and stony soils and can be distinguished quite easily by a few characteristics, including the fruits, which consist of a few very glaucous drupes which remain enclosed in large persistently erect sepals. Found only on the banks of the Royal Canal, '81, DD.

Rosa canina   Dog Rose

— 2 — — — — — — — — — 12 13 14

Native, occasional.

This species includes both the native dog rose as well as plants that have arisen from the wild rootstock of discarded cultivated roses. The two records which are most likely to be native stock are: railway embankment off St Attracta Road (12), '80, PWJ and in a large waste ground at Montpelier Gardens (13), '80, DNFC.

Geum urbanum   Wood Avens

— — — 4 — — — — — — 11 — 13 —

Native, rare.

A hedgerow and woodland species. This plant is probably a relict species in Dublin. It is difficult to see why it would have been cultivated or how it could have been introduced accidentally. It occurs in the grounds of St Brendan's Hospital, under trees (11), '81, DD, JRA, and in a derelict garden of an old house off Arbour Hill, also under trees. Growing with *Galium odoratum* (13), '80, DNFC. Waste ground off Charlemont Street (4), '82, PWJ.

Potentilla anserina   Silverweed

1 2 — 4 5 6 7 8 9 — 11 12 13 14

Native, very common.

Found in ruderal habitats in many waste grounds throughout the city. It spreads easily by stolons and flourishes in open habitats.

86

P. reptans   Creeping Cinquefoil
1 − 3 4 − 6 7 − − − 11 12 13 14
Native, frequent.
This plant occurs usually in a closed grassy plant community in the inner city. It is sometimes a weed of poorly
kept lawns.

Alchemilla mollis   Lady's Mantle
− − 3 − − − − − − − − − − −
Introduced, very rare.
A garden escape found in a single station in the city. It
was growing on a heap of soil and was obviously self-
sown. Earlsfort Terrace waste ground, '81, PWJ, JRA.

Aphanes arvensis   Parsley-piert
− − − − − − − − − − − 12 − −
Native, very rare.
Very rare in Dublin probably because it grows on limestone walls or a good moss-covered sandy or gravelly substrate. In County Dublin it is recorded in tilled fields and
roadsides. Railway Embankment near Fassaugh Road, '81,
PR.

Malus domestica   Apple
1 2 − − − − 7 8 − − − − − −
Introduced, occasional.
Grown very commonly in suburban gardens, but rarely in
the inner city. A few trees have become established in the
city, probably self-sown from apple-cores.

Cotoneaster microphyllus   Cotoneaster
− − 3 − − − − − − − − − − 14
Introduced, rare.
This plant is occasionally self-sown from seeds dispersed
by birds. The red berries are popular with garden birds
and this species is commonly grown. Edge of Royal
Canal near Cross Guns Bridge (14), '81, DNFC.

Crataegus monogyna   Hawthorn
  − 2 − 4 − − − 8 − − − 12 − 14
Native, frequent.
Grows in hedges and some waste places. It is also grown in some gardens. The red berries are readily dispersed by birds.

LEGUMINOSAE

Laburnum anagyroides   Laburnum
  − − 3 − 5 − 7 − − − 11 − − −
Introduced, occasional.
Seedlings found whenever a mature tree is growing nearby. Occasionally young plants are found in the absence of a parent. Grand Canal Harbour (7), '81, DNFC. Seedlings growing under a mature tree in Wilton Place Square (3), '80, KD, PWJ.

Ulex europaeus   Furze, Gorse
  1 − − − − − − − − − 11 − − −
Native, rare.
It invades poorly-tended, slightly acid soils, establishing itself by seed. St Brendan's Hospital (11), '81, MSS, PWJ. Grand Canal Dock (1), '79, PWJ.

Cicer arietinum   Chickpea
  − − − − − − − − 9 − − − − −
Introduced, very rare.
A casual only, found in Dublin in a single station at Mountjoy Square, growing outside a health food shop, with *Lens culinaris*. '79, JRA. Chickpeas have never been grown as a crop in Ireland, nor commonly eaten.

Vicia cracca   Tufted Vetch
  − − − − − − − − − − − − − 14
Native, rare.
This plant prefers the stable habitat of hedgerows or fallow fields to the waste ground environment. Found in the inner city on the banks of the Royal Canal.

V. sepium   Bush Vetch
– – – 4 – – – – – – – – 12 – –
Native, locally frequent.
This plant is not common in the inner city due to a
shortage of suitable habitats. It likes a grassy habitat and
is often found in hedges.

V. sativa   Common Vetch
– – – 4 – – 7 – – – – 12 13 –
Native, occasional.
Prefers a sandy or gravelly habitat. Most waste grounds
are not suitable for this plant.

V. faba   Broad Bean
1 – – – – – – – – – – – – – –
Introduced, rare.
Found only as a discard with household rubbish. It does
not become established and is merely a casual.

Lens culinaris   Lentil
– – – – – – – – 9 – – – – –
Introduced, very rare.
A casual only, found growing with *Cicer arietinum* outside a health food shop off Mountjoy Square. It is not a
cultivated crop in Ireland, '79, JRA.

Lathyrus pratensis   Meadow Vetchling
– – – – – – – – – – – – – 14
Native, rare.
Prefers a grassy habitat and is rare in the inner city. Part
of a limestone grassland community on the banks of a
railway line on the Royal Canal bank, '81, DNFC.

Pisum sativum   Pea
1 2 – – – – – – – – – – – – –
Introduced, rare.
Rarely becomes established and does not persist from
year to year. It only occurs as a discard with household
rubbish.

1. Stellaria media
2. Cerastium fontanum
3. Trifolium dubium
4. Medicago lupulina
5. Mecurialis annua
6. Euphorbia peplus

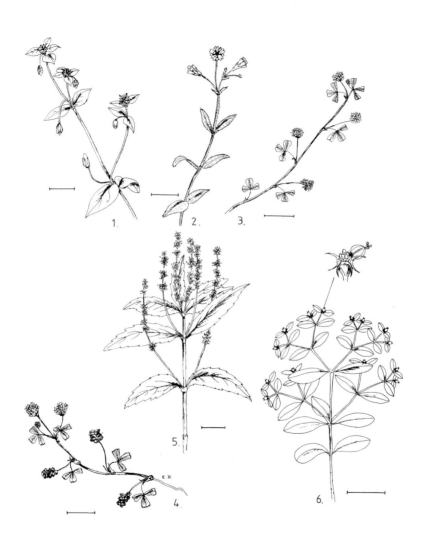

1. 2. 3.

5.

4.

6.

Melilotus altissima   Melilot
$$- - - - - - 7\ 8\ - - 11\ - - -$$
Introduced, rare.
A rare plant found in only a few localities in the city.
Grows in waste grounds in open plant communities but
does not seem to spread much within the city, even
though in some localities a considerable number of plants
are found which produce a lot of apparently good seed.
It was a herb plant in the past, used in homeopathy. Its
presence in Dublin is probably as a relic of its former use
in this way.

Medicago lupulina   Black Medick
$$1\ - 3\ 4\ 5\ 6\ 7\ 8\ 9\ 10\ - 12\ 13\ 14$$
Occurs in open waste grounds. It does not appear to be
successful in competition with grass species. It is one of
the species that is worn as shamrock on St Patrick's
Day. (Figure p. 91)

Trifolium repens   White Clover
$$1\ 2\ 3\ 4\ 5\ 6\ 7\ 8\ 9\ 10\ 11\ 12\ 13\ 14$$
Native, abundant.
Occurs in virtually every waste ground and ruderal site.
Also on lawns and in grasslands. It spreads easily by
stolons and is frequent in both open and closed plant
communities.

T. hybridum   Alsike Clover
$$- - - - - - - - - - - - - 14$$
Introduced, rare.
It is occasional in Ireland as a relic of cultivation. Found
on the banks of the Royal Canal, '80, HH.

T. campestre   Hop Trefoil
$$- - - - - - - - - - - - - 14$$
Native, rare.
A plant of dry pastures and waste places. It was found
growing on the Royal Canal bank, '80, HH.

T. dubium   Yellow Clover
   1 – 3 – – 6 – – 9 – 11 12 13 14
   Native, very frequent.
   Frequently encountered in waste grounds throughout the
   city. More common in long-established sites. It is less
   frequent than *Medicago lupulina* with which it is often
   confused. This species is the most commonly worn shamrock on St Patrick's Day. (Figure p. 91)

T. pratense   Red Clover
   1 2 3 4 5 6 7 8 9 10 11 12 13 14
   Native, very common.
   Occurs in similar places to *T. repens* but less abundantly.

Lotus corniculatus   Bird's Foot Trefoil
   – – – – – – – – – – – 12 – 14
   Native, rare.
   It grows as part of a limestone grassland community,
   occurring only on the railway bank near St Attracta Road
   in a dry sandy habitat (12), '80, PWJ; and on the junction
   between the Royal Canal bank and a railway embankment
   near Cross Guns Bridge, '80, DNFC.

Anthyllis vulneraria   Kidney Vetch
   – – – – – – – – – – – – 14
   Native, very rare.
   Grows on a railway embankment with an interesting limestone grassland flora near Cross Guns Bridge. This species
   is a calcicole of sandy and gravelly habitats, '80, DNFC.

                       OXALIDACEAE

Oxalis corymbosa
   – – – – – 6 – – – – 11 – 13 14
   Introduced, occasional.
   This common garden plant with pink flowers quite frequently escapes into waste grounds and onto roadsides.

Geranium molle
1 — — — — — — — — — — — — —
Native, rare.
Waste ground off Sir John Rogerson's Quay, '82, PWJ.

G. dissectum   Cut-leaved Crane's Bill
— — — — — — 7 8 — — — — 13 —
Native, occasional.
Waste grounds and roadsides. It is not a primary coloniser of waste ground, but prefers open grassland communities which are infrequent in the city.

G. robertianum   Herb Robert
— — 3 4 — — 7 8 — — 11 12 13 —
Native, frequent.
Growing on walls and in waste grounds. A biennial, this species can colonise open ground, but seed dispersal is not very far-reaching. This plant is probably a relic from when hedges and soil-covered walls were more numerous in or near the city centre.

Erodium moschatum   Storksbill
— — — 4 — — — — — — — — — —
Introduced, very rare.
This seaside weed is rare in Ireland. Its only station in the inner city is in an unlikely locality, numerous small plants being found in a lawn outside the Arts Building, Trinity College. Its mode of arrival is unknown, but it may have been introduced with sand spread over the newly laid lawn, '81, JRA.

Tropaeolum majus   Garden 'Nasturtium'
— 2 — — — — — — — — — — — —
Introduced, rare.
This is a common garden plant. It is often seen growing

rampant in gardens, but is rarely naturalised in the inner city. The single station on the Grand Canal near Dolphin's Barn Bridge was the only convincingly naturalised locality, though there was some evidence that the site had been used as a dumping ground for garden waste, '81, MSS, PWJ.

LINACEAE

Linum usitatissimum   Garden Flax
— — — — — — 7 — — — — — — —
Introduced, very rare.
This is the cultivated flax but does not seem to have occurred in Dublin in the past. It is also a constituent of birdseed which is probably its source in Dublin. A single station for this plant was found on the filled-in Grand Canal Harbour, '81, DNFC.

EUPHORBIACEAE

Mercurialis annua   Annual Dog's Mercury
1 2 3 4 5 6 7 8 9 10 11 12 13 14
Introduced, abundant.
One of the commonest and most abundant weeds of the inner city. It is found in most ruderal habitats from waste grounds and roadsides to gardens and walls. Never common outside towns of the east and south, Dublin remains its headquarters, as it was in Colgan's time (1904). Male and female flowers are on separate plants. (Figure p. 91)

Euphorbia helioscopia   Sun Spurge
— — 3 4 — 6 — — — — 11 — 13 —
Possibly introduced, occasional.
A weed of cultivated and recently disturbed ground. Much less frequent than the next species, although it grows in a similar type of habitat. It may have a preference for more nutrient-rich soils. (Figure p. 69)

E. peplus   Petty Spurge
1 2 3 4 5 6 7 8 9 10 11 12 13 14
Possibly introduced, common.
It is common on recently disturbed and cultivated
ground, but it tends to disappear when the plant community closes and it succumbs to competition. (Figure
p. 91)

ACERACEAE

Acer pseudoplatanus   Sycamore
1 2 3 4 5 6 7 8 9 10 11 12 13 14
Introduced, abundant.
This is probably the commonest tree in the inner city.
It seeds itself freely in many habitats and grows quickly.
It is probably planted much less frequently than it is
self-sown.

A. macrophyllum   Oregon Maple
– – – 4 – – – – – – – – – – –
Introduced, very rare.
Seedlings of this species are quite common in Trinity
College, near and beneath two large parents. They are
usually weeded out before they attain any size.

HIPPOCASTANACEAE

Aesculus hippocastanum   Horse-Chestnut
– – 3 4 – – – – – – – – – –
Introduced, rare.
Commonly planted throughout the city though apparently rarely self-sown in Dublin. Most trees seen
were probably planted and few seedlings were found. The
large 'conkers' probably need a good depth of soft soil to
germinate and develop into saplings, as elsewhere in the
city seedlings are frequent.

96

BALSAMINACEAE

Impatiens parviflora
– – – 4 – – – – – – – – – –
Introduced, very rare.
Found only in a single station, behind the Biochemistry
Building in Trinity College. It may be a discard from
teaching material. Although abundant in this locality, it
probably will not spread due to its preference for a shady
site and its annual habit, '80, MSS.

I. glandulifera   Balsam
– – 3 4 – – 7 – – 10 – – – –
Introduced, occasional.
A garden escape that is occasional in parts of the city,
generally on waste grounds. It prefers a moist habitat but
surprisingly has not spread along either canal. Large
populations were found in the vicinity of Charlemont
Street (4), '80, PWJ.

AQUIFOLIACEAE

Ilex aquifolium   Holly
– – 3 4 – – – – – – – – – 14
Native, rare.
Although commonly planted in gardens, it is rarely found
in waste grounds or ruderal sites. Wall top, College Lane,
Trinity College, no doubt from a seed dropped by a bird
(4), '80, PWJ.

VITACEAE

Parthenocissus quinquefolia   'Virginia Creeper'
– – – – – – – 8 – 10 11 – 13 –
Introduced, occasional.
A garden plant that is found occasionally in the city in
waste grounds. Whether it is naturalised or not is not
clear however, as it has a great ability to survive in a
habitat long after the wall on which it was planted is gone

and even after all traces of the garden where it was have disappeared. It is rampant in a number of waste grounds in the city. It also grows on some walls where it has certainly been planted. It can be distinguished from the next species by its tendrils with (3-)5-8(-12) branches, the tendril ending in an adhesive disc.

P. inserta   'Virginia Creeper'
$- - 3\ 4 - 6 - - 9 - - 12 - -$
Introduced, occasional.
Very similar to *P. quinquefolia* in growth and habitat but occurs more frequently. It can be distinguished from the last by having tendrils with 2-5 branches, often swollen at the apex but without an adhesive disc.

MALVACEAE

Malva sylvestris   Mallow
$1\ 2\ 3\ 4 - 6\ 7\ 8\ 9\ 10\ 11 - - 14$
Probably introduced, common.
Common in many parts of the city in waste grounds, occurring in both open and closed plant communities. A perennial plant, it can attain 1 m in height.

M. neglecta
$- - - - - - - - - - - - - 14$
Introduced, very rare.
It can be distinguished from the last species by having prostrate stems and smaller flowers, 2 cm across, as opposed to 3-4 cm. Recorded from the bank of the Royal Canal, '80, HH.

THYMELAEACEAE

Daphne laureola
$- - - - - - - - - - - - - 14$
Introduced, very rare.
A garden plant that has become naturalised on one railway embankment near Cross Guns Bridge, on the Royal Canal bank, '81, DNFC.

GUTTIFERAE

Hypericum calycinum   Rose of Sharon
  − − 3 4 − 6 − − − − 11 − 13 −
Introduced, frequent.
A garden plant that shows a great ability to spread and
become naturalised in waste grounds, whenever they occur
near gardens. A conspicuous plant, with large yellow
flowers.

H. androsaemum   Tutsan
  − − − − − − − − − − − − 13 −
Native, very rare.
This is a shade-loving plant that was probably reduced in
frequency when trees and hedges gave way to urban
settlement. Only one station for the plant, in the grounds
of St Brendan's Hospital, under trees, '81, DD, JRA.

H. pulchrum
  − − − 4 − − − − − − − − − −
Native, very rare.
A calcifuge, found in a single station only, in a flower-pot
in Dawson Street, '81, MS.

H. perforatum
  − − − − − − − 8 − − − − − −
Native, very rare.
Not a good coloniser of waste ground, although Colgan
notes that it is frequent on stony ground and wall-tops.
Waste ground near Sheriff Street and the Royal Canal,
'81, PWJ, JRA, DD.

ONAGRACEAE

Fuchsia magellanica   Fuchsia
  − 2 − − − − − − − − − − 13 −
Introduced, occasional.
A garden plant that is extensively planted in rural Ireland,
especially as hedges but also in gardens. It is commoner in

Dublin than is indicated above, but is probably planted in all but a few cases.

Circaea lutetiana   Enchanter's Nightshade
$-- 3\ 4 ---- --- 11 ---$
Native, rare.
A weed of old and shady gardens. It prefers a moist base-rich soil. Near Botany Department, Trinity College (4), '79, PWJ. Yard off Lad Lane (3), '80, PWJ, KD. Grounds of St Brendan's Hospital (11), '81, MSS, PWJ.

Oenothera biennis   Evening Primrose
$--- - 5 -- ------ 13 -$
Introduced, rare.
This plant is a garden escape and widely naturalised in Europe. There are at least a dozen species found in Europe and all are of American origin. Taxonomically they are very difficult and some species are reputed to have arisen in Europe from American parentage. Unfortunately this genus is not common in Dublin waste grounds as it is certainly one of our most striking weeds. It is only found close to the Liffey in Dublin city.

Epilobium angustifolium   Rosebay Willow-herb, Fireweed
1 — 3 4 5 6 7 8 9 10 11 12 13 14
Native, common.
A coloniser of many Dublin waste grounds, it is frequently cultivated in gardens and escapes to colonise ruderal habitats. Although it is native in Ireland, it only occurs rarely in some mountain habitats and is certainly intro-duced into Dublin. It occurs in waste grounds that are left undisturbed for a few years. Apparently absent from the banks of the Grand Canal. (Figure p. 14)

E. hirsutum   Hairy Willow-herb
1 2 3 4 5 6 7 8 9 10 11 12 13 14
Native, common.
This weed of damp waste grounds is common in Dublin and widespread but is never as abundant as the previous

species. Species appear to fluctuate greatly in numbers from year to year for some unexplained reason. An attractive white-flowered form has been noted in a number of places in the city; Grand Canal bank by Grand Canal Street Bridge (1), '81, PWJ, JRA, and waste ground at Grangegorman (11), '81, MSS, PWJ.

E. parviflorum   Small-flowered Willow-herb
1 – 3 – – 6 7 – – – – 12 – 14
Native, occasional.
One of the rarer species of willow-herb to be found in Dublin. It is usually found as scattered individuals and never occurs in large populations. It is easily recognised by its hairy appearance like *E. hirsutum*, but it is smaller and has smaller flowers. A hybrid between these two species has been reported from zone 1 in 1979 (PHC), but cannot be determined with complete certainty.

E. montanum   Broad-leaved Willow-herb
1 2 3 4 5 6 7 8 9 10 11 12 13 14
Native, very common.
Common in all waste grounds and gardens. It can be distinguished from *E. obscurum* and *E. adenocaulon*, which it resembles closely, by its 4-lobed stigma and larger flowers. The two other species have a club-shaped stigma and flowers 6-8 mm across (see below).

E. obscurum
1 2 3 4 5 6 7 8 9 10 11 12 13 14
Native, very common.
Common in waste grounds and gardens. It is not just confined to damp habitats in Dublin, but is less common on walls. It may be declining somewhat in the city due to pressure from the new arrival, *E. adenocaulon*. This may in part be an artificial decline however, as some of the previous records for the species are probably mis-identifications of the latter species. Despite this it was still found in all 14 zones during 1981.

E. adenocaulon

1 2 3 4 5 6 7 8 9 10 11 12 13 14

Introduced, common.

Occurs commonly in waste grounds, on walls and at their bases. A recent arrival in Ireland, it was first noticed in Dublin in the beginning of 1981, the last year for field-work for this project. However, it was subsequently found in all 14 divisions of the city. The striking fact is that although it must have been present in the city before 1981, it was not common and may have been overlooked. It is certainly spreading very quickly through the city. It was first noticed as a weed at the National Botanic Gardens at Glasnevin, outside the limits of this study. It is also now a common weed at the Phoenix Park Garden Centre (DD). Both of these places could well act as centres for dispersal in Dublin. It occurs in both wet and dry habitats and seems to be replacing *E. obscurum* as the commonest city willow-herb. It can be distinguished easily from the other species by the presence of sticky glandular hairs on the stem.

A few specimens similar to this plant and to *E. obscurum* were noticed in the city also. They differed only in that they were devoid of glandular hairs throughout the plant. They seem to be referable to *E. tetragonum*, a plant as yet unrecorded in Ireland. Further work will however be necessary to determine their status.

E. nerterioides   New Zealand Willow-herb

— — — — — — — — — — — — 14

Introduced, very rare.

A New Zealand species, now widespread in many of the Irish mountains. It was originally introduced as a rock-garden plant. Its preference for a damp habitat is pro-bably the only limiting factor in its spread in Dublin where it only has a single station. On Cross Guns Bridge, over the Royal Canal, '81, DD.

Hybrids between many of these Willow-herb species are occasionally found and may be difficult to identify with

102

certainty. There is still much valuable work to be done on the genus in the city, especially with regard to the *E. obscurum*, *E. adenocaulon* and *E. tetragonum* group, to determine fully their status.

HIPPURIDACEAE

Hippuris vulgaris   Mare's Tail
— — — — — — — — — — — — — 14
Native, rare.
A plant of ditches, pools and lake-margins that in the inner city occurs only in the Royal Canal and on its banks, '80, HH.

ARALIACEAE

Hedera helix   Ivy
— 2 3 4 5 6 7 8 — 10 11 12 13 14
Native, frequent.
This plant is generally found clothing old walls around the city. Its berries are popular with birds and it sows itself and spreads easily. Thus it has probably always been common in and around the city.

UMBELLIFERAE

Anthriscus sylvestris   Cow-parsley
— — — — 5 — — — — — — 12 — —
Native, rare.
Occurs rarely in the inner city in waste grounds and shady places. A hedgerow plant in the country.

Smyrnium olusatrum   Alexanders
— 2 — — — 6 — — — — — — — 14
Introduced, rare.
This plant has a distinctly coastal distribution in Ireland where it grows on roadsides and in waste places. Colgan (1904) notes its cultivation in the eighteenth century as a vegetable like celery and regarded it as a relic of cultiva-

tion. In the city it is found only close to or on the Grand
and Royal Canal banks.

Aegopodium podagraria   Bishop's Weed, Goutweed
– – – – – – – – – – 10 – – – –
Introduced, rare.
Waste grounds. Probably introduced during the Middle
Ages as a medicinal pot herb. It is a troublesome weed
of gardens, but rare in the inner city.

Oenanthe crocata   Water-dropwort
– 2 – – – – – – – – – – 14
Native, rare.
A plant of ditches and river-sides. Found only by the
Grand and Royal Canals in the city.

Aethusa cynapium   Fool's Parsley
1 – 3 – – 6 7 8 9 10 11 – 13 14
Introduced, frequent.
Quite a common weed of cultivated ground and waste
places, preferring nutrient-rich soils. It is native in Britain.
(Figure p. 139)

Foeniculum vulgare   Fennel
– – – – – – – – – – – – 14
Introduced, very rare.
It was widely naturalised in the Phoenix Park region in
1904. A relic of cultivation. It has a single station in the
inner city, on the Royal Canal bank, west of Cross Guns
Bridge, '80, HH.

Conium maculatum   Hemlock
– – – – – – 7 – – – – – – –
Possibly introduced, rare.
This plant seems to have been more common in 1904,
growing near dwellings or old ruins. It is very poisonous
and has a single station in the inner city on a waste ground
near Dolphin's Barn.

104

Apium graveolens   Celery

$- - - - 5 - - - - - - - - - -$

Native, very rare.

Although this species is native in Ireland, this record is
for cultivated celery that has become established in a
waste ground. The cultivated variety is var. *dulce*. The
native species is a plant of salt marshes and brackish
ditches. Waste ground near St Patrick's Cathedral, '80,
MSS, PWJ.

A. nodiflorum   Fool's Water-Cress

$- 2 - - - - - - - - - - - - 14$

Native, locally frequent.

This plant is restricted to the canals where it is locally
abundant. It resembles water-cress and is sometimes
eaten.

Angelica sylvestris   Wild Angelica

$1 \ 2 \ 3 \ 4 - - 7 \ 8 - - - 12 - 14$

Native, frequent.

Occurs commonly in damp waste grounds and on the
canal banks. It also occurs in shady places.

Pastinaca sativa   Wild Parsnip

$- - - - 5 - - - - - - - - - -$

Introduced, very rare.

In Ireland this plant is an escape from cultivation. There
is only a single station for the plant in the inner city, in
a waste ground near St Patrick's Cathedral, '80, MSS,
PWJ.

Heracleum sphondylium   Hogweed

$1 \ 2 \ 3 \ 4 - 6 \ 7 \ 8 - 10 - 12 \ 13 \ 14$

Native, common.

Found commonly in waste grounds and ruderal habitats.
It is a common weed of roadsides and other grassy places.
(Figure p. 139)

H. mantegazzianum
Giant Hogweed

$- - - - - - - - - - 11 - - -$

Introduced, very rare.

It is a plant of damp habitats and is found in abundance on the banks of the river Tolka, outside the limits of the inner city, at Glasnevin. There is a single station for this striking plant in the city in the grounds of St Brendan's Hospital, '81, DD, JRA.

Daucus carota
Wild Carrot

$- - - - - - 7 - - 10 - 12 - 14$

Native, occasional.

Although this is a native plant in Ireland three of the records for the species may well be the cultivated carrot, subsp. *sativus* (7, 10 and 12). The record for the plant on the banks of the Royal Canal in limestone grassland is the most likely to be of native stock. It is a plant of waste places, dry banks and grassland.

PRIMULACEAE

Primula vulgaris
Primrose

$- - - - - - - - - - - 12 - -$

Native, very rare.

A plant of shady banks and woods, it has a single locality in the inner city. It was found growing on the walls of a disused railway line, near Cabra Road, '81, PR, PWJ. There are very few suitable habitats in the city so it is not surprising that it is rare.

P. veris
Cowslip

$- - - - - - - - - - - - 13 -$

Native, very rare.

This is an unexpected plant in the flora. It is found in poor pastures particularly inland in Ireland. The site it occupied contained a number of other interesting and unexpected plants and may thus be a piece of relic grassland. Waste ground at Montpelier Hill, '81, DNFC.

106

Lysimachia nummularia   Moneywort, Creeping Jenny
– – 3 – – – – – – – – – – – – –
Native, very rare.
Although a native Irish plant it is certainly an escape
from cultivation in this instance. It is found in the wild in
damp grassy places, lakeshores and riversides. In this case
however it was growing on a grassy roadside patch, near
Wilton Place, '80, KD, PWJ.

Anagallis arvensis   Scarlet Pimpernel
– – – 4 – – – – – – 11 – 13 –
Native, rather rare.
This species is a weed of tilled fields, waste grounds and
cultivated places. It is odd that it is so rare in Dublin city,
however it seems to be more a weed of rural tillage than a
city waste ground weed. All material seen in Dublin has
scarlet flowers. Plants, at coastal sites especially, frequently
have flesh-coloured flowers in Ireland.

OLEACEAE

Fraxinus excelsior   Ash
– 2 3 4 5 6 7 8 9 10 11 12 13 14
Native, common.
Common in most districts. It seeds itself freely and quite
a number of apparently self-sown specimens have been
found as mature or semi-mature trees. It was not found
in zone 1, probably because of the absence of any parent
trees to provide seed. Ash is quite commonly planted in
many city gardens.

Ligustrum ovalifolium   Japanese Privet
– – 3 – – 6 – – – – 11 12 13 –
Introduced, occasional.
A garden plant that is very commonly grown in Dublin.
It is the principal hedging plant used in the city, often as
the golden-leaved cultivar. It may become locally naturalised and the records above are for sites where it is
unlikely to have been planted.

107

GENTIANACEAE

Blackstonia perfoliata   Yellow-wort
$— — — — — — — — — — — 12 — —$

Native, very rare.
In Ireland it is a plant of calcareous gravelly banks, sand-hills and rocky places. The railway embankment where it was found is a similar habitat to its natural one. Railway embankment near St Attracta Road, '80, PWJ.

RUBIACEAE

Galium odoratum   Woodruff
$— — — — — — — — — — — — 13 —$

Native, very rare.
In one locality with *Geum urbanum* in an old garden, amongst trees. These plants may be relics of a woodland habitat. Off Arbour Hill, '80, DNFC.

G. palustre   Marsh Bedstraw
$— 2 — — — — — — — — — — — 14$

Native, rare.
An inhabitant of ditches and marshy places. It grows on the Grand and Royal Canal banks only in the inner city.

G. verum   Lady's Bedstraw
$— — — — — — — — — — — — 14$

Native, rare.
This species occurs on a railway embankment beside the Royal Canal near Cross Guns Bridge, with other lime-stone grassland species. '80, DNFC.

G. aparine   Robin-run-the-hedge, Goosegrass
1 2 3 4 5 6 7 8 9 10 11 12 13 14

Native, abundant.
Abundant in most city habitats. It is one of the most widespread and common city weeds. Its fruit, covered with hooked bristles, facilitates its rapid spread into a new waste ground. It has a preference for nutrient-rich soils.
(Figure p. 111)

CONVOLVULACEAE

Calystegia sepium   Bindweed
1 2 3 4 — 6 7 — 9 10 11 12 13 14
Native, common.
This weed of gardens, waste places and roadsides is widespread in Dublin. It is troublesome in gardens and is
difficult to eradicate once established, as it regenerates
from tiny fragments of root left in the soil. (Figure
p. 111)

C. sylvestris   Bindweed
— 2 — — — — 7 — — — 11 — — 14
Introduced, frequent.
Much less common than the last species, but frequent
nevertheless and probably more than the records indicate.
The two species can be distinguished reliably only on
floral characteristics. *C. sepium* has a corolla about 50 mm
long and flat, or slightly inflated bracts. *C. sylvestris* has a
corolla 65-75 mm long and strongly inflated bracts.

Convolvulus arvensis   Small Bindweed
1 — 3 4 — 6 7 — — — 11 12 13 14
Native, common.
This plant occurs in tilled and cultivated ground and
waste places. It is an extremely persistent and noxious
weed and is commonest on light basic soils. The first
county record for the species was 'Upon the mudwalls in
Cabera-Lane' (Threlkeld 1726). A very early record for
zone 12!

BORAGINACEAE

Pentaglottis sempervirens
— 2 — — — — — — — — — — — —
Introduced, very rare.
A garden plant that rarely escapes to become established
in the wild. On the bank of the Grand Canal near Leeson
Street Bridge, '79, MSS.

1. Galium aparine
2. Calystegia sepium
3. Cymbalaria muralis
4. Chamomilla suaveolens
5. Tanacetum parthenium

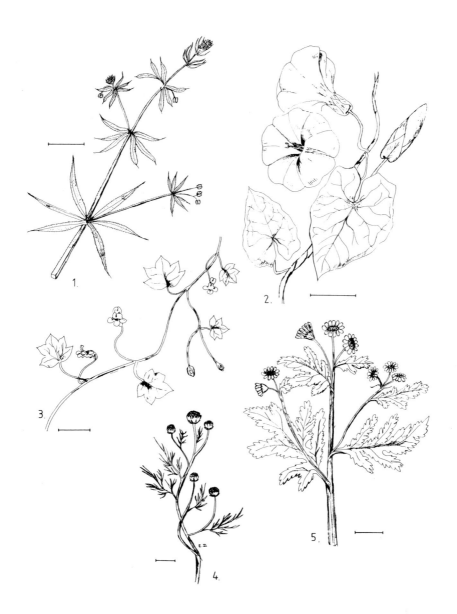

1.

2.

3.

4.

5.

E.II

111

Myosotis arvensis   Forget-me-not
— — — 4 — 6 — — — — 11 — 13 14
Native, occasional.
A weed of dry waste places that is occasionally encountered throughout the city.

CALLITRICHACEAE

Callitriche spp.   Water Starwort
1 2 — — — — — — — — — — — —
Native, locally frequent.
This genus occurs quite widely in the Grand Canal. It is very difficult taxonomically, but some specimens determined as *C. brutia* were collected in the Grand Canal Dock, near Grand Canal Street Bridge, (1), '81, PWJ, JRA and west of the bridge, (2), '81, PWJ, JRA. Other *Callitriche* specimens were spotted in the section between Leeson Street Bridge and Charlemont Bridge, (2), '81, MSS. These were not determined to the specific level. Shackleton (1976) gives a tentative identification of *C. platycarpa* for specimens found near the Parnell Bridge on the canal.

LABIATAE

Galeopsis tetrahit   Hempnettle
— — — — — — — 8 — — — — — —
Probably introduced, very rare.
A rare plant found in only a single city station, close to the Jetfoil terminus, on the North Wall Quay. It was found in a new flower-bed beside cattle pens. Cattle or trucks may have brought in the seed from its more usual habitat of tillage ground in peaty upland areas of County Dublin. '81, PWJ, JRA, DD, EL.

Lamium purpureum   Red Deadnettle
— 2 — 4 — 6 7 — — — — — 14
Native, occasional.
A weed of cultivated plots and tilled fields, preferring

112

disturbed ground. It is not very common in Dublin city. Allotments by the Grand Canal, near Dolphin's Barn Bridge, '81, MSS, PWJ.

Ballota nigra   Black Horehound
— — — — — — — — — — — — — 14
Introduced, very rare.
It is a plant found mainly in the east of Ireland growing in waste places and on roadsides. Grown in cottage gardens, Colgan notes its survival for over a century in some sites where it has escaped. Found on the Royal Canal banks only, in the inner city, '80, HH.

Stachys sylvatica   Hedge Woundwort
— 2 — — — — — — — — — 13 —
Native, rare.
This plant of shady places and hedges occurs only rarely in Dublin city. It has been found in a large waste ground near Montpelier Hill, (13), '81, DNFC, and in one locality on the Grand Canal bank, (2), '80, MSS.

S. palustris   Marsh Woundwort
— 2 — 4 — — — — — — — — 14
Native, rare.
A weed of tilled fields, ditches and river-banks. It is rather rare in Dublin city, occurring on the banks of the two canals and in a new flower-bed outside the Botany School, Trinity College, (4), '81, DD, PWJ.

Glechoma hederacea   Ground Ivy
— — — — — — — — — — 11 — — —
Native, very rare.
A shade loving plant of hedgerows and woodland. This species, growing with other woodland species are probably relics of this habitat on a site which has never been cleared or built on. Grounds of St Brendan's Hospital, '81, DD, JRA.

Prunella vulgaris   Self-heal

1 — 3 — — — — — — — — 12 13 —

Native, occasional.

This species usually occurs in long-established open grassy communities. It is not generally a primary coloniser of waste ground.

Origanum vulgare   Marjoram

— — — — — — 7 — — — — — — —

Native, very rare.

This plant is a limestone grassland plant in Ireland found in a single locality in the inner city. It has not been found on the banks of the Royal Canal although Colgan recorded it there and on the Tolka River. His stations may, however, be outside the inner city limits. It is probably a garden escape in its station in a waste ground near Dolphin's Barn, '80, MH.

Lycopus europaeus   Gipsywort

1 2 — — — — — — — — — — — 14

Native, occasional.

A plant of river margins and ditches. It only occurs by the two canals in the city, where it is quite frequent.

Mentha aquatica   Water Mint

— 2 — — — — — — — — — — — 14

Native, occasional.

Grows quite frequently on the wet margins of the two canals.

M. suaveolens (*M. rotundifolia*)   Mint

— 2 — — — — — — 9 — — — 13 —

Introduced, rather rare.

A garden escape found in three localities in the city. Waste ground off Montpelier Hill, (13), '80, DNFC. Bank of the Grand Canal, (2), '81, MSS. Waste ground near Mountjoy Square, (9), '80, DNFC.

Solanum dulcamara   Bittersweet
   1 2 3 − − 6 7 8 9 − − 12 13 14
   Native, frequent.
   This plant occurs very frequently in waste grounds and
   shady places. It is commonest in shady derelict gardens,
   but does occur in open habitats in waste grounds, particularly those which have been established for a number
   of years. The berries it bears are poisonous.

S. tuberosum   Potato
   1 2 3 4 5 6 7 − 9 10 11 12 13 −
   Introduced, common.
   This common vegetable is often discarded in waste places
   and soon becomes established. Although in some places
   it survives from year to year, it is probably maintained in
   many others by new discards. It is commonest near houses
   and was not found on the Royal Canal bank.

Lycopersicon esculentum   Tomato
   1 2 3 − 5 6 7 8 9 10 11 − 13 −
   Introduced, common.
   The ripe fruits of tomatoes are frequent discards from
   households and fruit markets. Plants very rarely reach
   maturity and behave only as casuals. Found in waste
   places and other ruderal sites.

BUDDLEJACEAE

Buddleja davidii   Buddleja, Butterfly Bush
   1 2 3 4 5 6 7 8 9 10 11 12 13 14
   Introduced, abundant.
   Having invaded Dublin relatively recently, this plant is
   now perhaps the commonest introduced weed of the city.
   In many waste grounds stands of mature *Buddleja* 'scrub'
   can be found. It grows rapidly and seeds itself freely in
   many habitats, growing in waste grounds, on walls, in the
   cracks of pavements and in gutters and drain-pipes. In

115

many cases wall plants of this species have attained a large size. Native to China, it was introduced to Britain late last century by Augustine Henry, a famous Irish pioneer of scientific forestry, and is now quite commonly planted in gardens. It is however a rare plant in grassy communities, apparently unable to withstand competition in its early stages of growth. (Figure p. 33)

SCROPHULARIACEAE

Verbascum thapsus   Mullein
1 − − 4 − − − − − − − − − − −
Native, rare.
Although this plant is native in Ireland, it is often planted in gardens and from there escapes onto waste grounds in the city. It is not common and has only been found in the inner city on two occasions. Waste ground off Charlemont Street, (4), '80, PWJ. Waste ground near the Grand Canal Dock, (1), '79, DNFC.

Scrophularia nodosa   Figwort
− − − − − − 7 − − − 11 − − −
Native, rare.
Found in two localities in the city. At the base of a wall, Queen's Street, (11), '81, PWJ, MSS. Waste ground near the Grand Canal Harbour, (7), '81, DNFC.

Antirrhinum majus   Snapdragon
1 2 3 4 5 6 7 8 9 10 11 12 13 −
Introduced, very common.
A common annual garden bedding plant that escapes frequently and becomes established on walls and piles of rubble. Although well naturalised in Dublin, populations are constantly being added to from garden seed. Most plants found are referable to subsp. *majus* or are probably hybrids. It is less common on deeper soils.

Linaria purpurea   Toad-flax
1 − 3 4 − − 7 − − − − − 13 −

116

Introduced, frequent.
This plant is locally frequent in the city, and found in waste grounds mainly south of the Liffey. It is quite commonly cultivated in gardens and from there becomes naturalised. Waste ground off Earlsfort Terrace, (3), '79, PWJ, DH. Waste ground off Charlemont Street, (4), '80, PWJ. Montpelier Hill, (13), '81, DD.

Cymbalaria muralis   Ivy-leaved Toadflax
1 2 3 4 5 6 7 8 9 10 11 12 13 14
Introduced, very common.
Grows almost exclusively on walls and at their bases. It was a very early rock garden introduction into Ireland. It is now well naturalised in Dublin and seldom cultivated in gardens. A white-flowered form is occasionally seen.
(Figure p. 111)

Digitalis purpurea   Foxglove
– – – 4 5 – – – – – – 12 13 –
Native, rather rare.
A common plant in most parts of Ireland, it is a calcifuge and therefore rather rare in Dublin. It is found generally on granite walls in the city.

Veronica serpyllifolia   Speedwell
– – – – – – – – – – – – 13 –
Native, very rare.
A plant of waste places, tilled ground, roadsides and pastures that has not invaded the city to any extent. Waste ground near Montpelier Hill in well-established grassland, '81, DNFC.

V. beccabunga   Brooklime
– – – – – – – – – – – 12 – –
A plant of wet habitats. Grows in the inner city in a disused railway line cutting near Cabra Road, '81, PR, PWJ.

V. anagallis-aquatica   Water-Speedwell
– – – – – – – – – – – 12 – –

Native, very rare.
A plant of wet and marshy habitats that occurs in a single inner city station on a disused railway line near Cabra Road, '81, PR, PWJ.

V. persica
    1 – – 4 – 6 7 – – – – – – –
Introduced, occasional.
Occurs occasionally on recently disturbed ground in waste grounds and gardens.

V. filiformis
    – – 3 – – – – – – – 11 – – –
Introduced, rather rare.
This little plant was originally cultivated, but is now a rather serious weed especially of lawns, though it is also found in waste places and on roadsides. Although common in suburban gardens, it is not common in the city due to its preference for moist habitats with closed vegetation.

Euphrasia officinalis agg.   Eyebright
    – – – – – – – – – – – – 13 –
Native, very rare.
A plant of this genus was noticed in one waste ground locality near Montpelier Hill in 1981 (DD). No specimen was collected and it was not determined to specific rank. On returning to re-examine the plant, it could not be located. This particular site contains other grassland plants uncharacteristic of city waste grounds.

Odontites verna   Red Bartsia
    – – – 4 – – – 8 – – – – 13 –
Native, occasional.
This is a weed of open plant communities that occurs occasionally in the city. No more than one or two individuals are ever found at any site. Mainly a weed of cultivation and pastures.

118

OROBANCHACEAE

Orobanche minor   Clover Broomrape
— — — — — — — — — — — — — — 14
Native, very rare.
This plant grows parasitically on the roots of clovers and
therefore lacks the green pigment, chlorophyll. It was
growing with other limestone grassland plants on a railway
embankment beside the Royal Canal near Cross Guns
Bridge, '80, DNFC.

PLANTAGINACEAE

Plantago major   Plantain
1 2 3 4 5 6 7 8 9 10 11 12 13 14
Native, abundant.
A common weed of waste grounds, roadsides, pavements and most ruderal habitats. There is a considerable
amount of variation in Dublin specimens, especially in
size. It grows best where the soil is deep and nutrient-
enriched.

P. lanceolata   Ribwort Plantain
1 2 3 4 5 6 7 8 9 10 11 12 13 14
Native, very common.
Not as abundant as the previous species but still very
common in grassy places and waste grounds and more
rarely on pavements and roadsides.

CAPRIFOLIACEAE

Sambucus nigra   Elder
1 2 3 4 — 6 7 8 9 10 11 12 13 14
Native, common.
Occurs on many waste grounds that are long-established
and in old derelict gardens. It is also frequent on the
banks of the canals and in hedges throughout the city.
Absent from zone 5. It often indicates nutrient-enriched
soils and is often planted in cottage gardens, seeding itself
freely.

119

Symphoricarpos rivularis   Snowberry

$- 2 - - - - - - - - - 12 - -$

Introduced, occasional.
Although this species is commonly planted in hedges
and gardens, it is not seen often in a semi-natural state.
It spreads freely by suckers.

VALERIANACEAE

Valeriana officinalis   Wild Valerian

$- 2 - - - - - - - - - - - 14$

Native, very rare.
A plant of damp ditches and stream-sides. Found only
on the banks of the Royal and Grand Canals, (2), '81,
MSS. West of Cross Guns Bridge, (14), '81, DD.

Centranthus ruber   Red Valerian

$1\ 2\ 3\ 4 - - - - - - 11\ 12\ 13\ 14$

Introduced, common.
This is quite a common plant on walls throughout much
of the city. It was originally introduced as a garden plant
but is very infrequently seen in gardens today. It prefers
old walls where it can become established in the crumbling
mortar. It is also found occasionally on old heaps of
rubble in waste grounds. A white flowered form is occasionally seen. The disjunct distribution of the plant in the
inner city is odd if one considers its abundance on walls
and on pavements around Trinity College and the gardens
of Fitzwilliam Square. It has been naturalised for a long
time and seeds itself freely. It is also abundant along the
railway lines in the city.

DIPSACACEAE

Dipsacus fullonum   Teasel

$- - - - - - - - 9 - - - - -$

Native, rare.
Although it is native in Ireland, it is commonly grown in
gardens and often escapes from there into the surrounding

120

waste places. It is especially prized for its use in dried flower arrangements. Not a common plant of city waste grounds however. It was formerly cultivated in North County Dublin to supply Dublin clothiers with carders for wool. Mountjoy Square, '81, DNFC.

CAMPANULACEAE

Campanula rapunculoides   Creeping Bellflower
— — — 4 — — — — — — — — — —
Introduced, rather rare.
A plant which is quite commonly grown in gardens and spreads into the surrounding area. Generally found in a dry habitat close to planted specimens in gardens. Only found once in the absence of any apparent source of seed from a garden. Waste ground near Charlemont Street, '80, PWJ.

COMPOSITAE

Eupatorium cannabinum   Hemp Agrimony
— — — — — — — — — — 11 12 13 —
Native, locally frequent.
Occurs by rivers, stream-sides and on damp rocky ground in Ireland. In the inner city it was found on the wall of the Liffey and in a few damp waste places. Waste ground near Montpelier Hill, (13), '80, DNFC. Wall of Liffey, Ellis Quay, (13), '81, PWJ. St Brendan's Hospital grounds, (11), '81, DD, JRA.

Bellis perennis   Daisy
1 2 3 4 5 6 7 8 9 10 11 12 13 14
Native, very common.
Usually in lawns. It also occurs on banks, roadsides and in pastures. The preferred habitat is one of closed plant communities with low grazing vegetation.

Aster novi-belgii   Michaelmas Daisy
— 2 — 4 — — — — — — — — — —

Introduced, rare.

A plant that is commonly grown in gardens and occasionally escapes. In Britain it is a more common escape, where it is frequent along railway lines. Waste ground near Charlemont Bridge (4), '80, PWJ. Well-established along the Grand Canal bank below Charlemont Bridge, (2), '81, MSS. It is grown in a garden on the south side of the canal a few hundred yards from these two sites from whence it probably escaped.

A. tripolium   Sea Aster, Sea Starwort
— — — — — — 7 — — — — — — 14

Native, rare.

A common salt marsh and sea-shore plant in Ireland. It occurs on mud at the base of the Liffey wall near St James's Gate, (7), '80, DNFC, and close to the mouth of the Royal Canal near Sheriff Street, (14), '81, JRA, DD, PWJ. It was growing with *Carex otrubae* in the latter station, in an apparently tidal part of the canal.

Galinsoga ciliata (*G. quadriradiata*)   Gallant Soldier
— — — 4 — — — — — — — — — —

Introduced, very rare.

Although this is now quite a common weed of gardens, arable land and waste places in Southern England and much of Europe, there are very few records of the plant in Ireland. The station recorded here is indeed the first confirmed record of the plant in the Dublin region. It was found growing in a flower-pot with bedding plants (*Lobelia erinus*) outside an hotel in Kildare Street, '80, PHC. It was still there in 1981 and the number of plants had increased to nine. In 1982 the plant had spread to Nassau Street.

Achillea millefolium   Yarrow
1 2 3 — — 6 7 8 — 10 11 12 13 14

Native, frequent.

Prefers a dry habitat. Found growing in pastures, on roadsides and waste places and occasionally on walls. This plant was used formerly in folk medicine in Ireland.

Matricaria maritima (*Tripleurospermum maritimum*) Mayweed
1 2 3 4 5 6 7 8 – – 11 12 13 14
Native, common.
A common weed of waste grounds in the city. It is difficult to distinguish from the next species except on seed characteristics. It is quite common around the docks but spreads readily inland growing on open ground. Colgan records it from all his Dublin regions. It is a semi-prostrate biennial or perennial with blunt fleshy leaf segments.

M. perforata
1 – – 4 – – – – 9 – 11 – – –
Introduced, occasional.
Found in similar habitats as the previous species. It is a more or less erect annual with apiculate non-fleshy leaf segments. The two species are separated with difficulty and this species has at times only been afforded subspecific rank.
Forms with oil-glands intermediate in shape between the two species grow in several waste grounds near the Liffey.

Chamomilla suaveolens   Pineapple-weed
1 2 3 4 5 6 7 8 9 10 11 12 13 14
Introduced, very common.
This plant occurs very commonly as a weed of waste places. It is especially frequent in trampled habitats such as grassy roadsides and tracks. It seems to be much commoner now than in Colgan's time. (Figure p. 111)

Chrysanthemum segetum   Corn Marigold
1 – – – – – – – – – – – – –
Introduced, very rare.
It is a plant of cultivated places in rural areas and is an unexpected plant in the inner city. Occurs in a waste ground near St Andrew's School, Pearse Street, '81, JRA, PWJ.

Tanacetum vulgare (*Chrysanthemum vulgare*)   Tansy
— — — — — — — 8 — — — — 13 —
Introduced, occasional.
Used to be extensively cultivated in gardens for ornamental purposes and as a herb. Although now grown infrequently in gardens, it is established in a few waste
grounds in the city. It has probably decreased in distribution since Colgan's time and is much less common
than the next species.

T. parthenium (*Chrysanthemum parthenium*)   Feverfew
1 2 3 4 5 6 7 8 9 10 11 — 13 14
Introduced, very common.
Very commonly naturalised in Dublin and occurring in
waste grounds, on roadsides and pavements. It was widely
cultivated in the past in gardens for its ornamental and
medicinal qualities, and to some extent still is. It seeds itself freely and is now well-established in Dublin. (Figure
p. 111)

Leucanthemum vulgare (*Chrysanthemum leucanthemum*)
                                                    Ox-eye Daisy
1 2 — — — — 7 — — — — 12 13 —
Native, occasional.
Occurs in dry grassland and meadows. Found in waste
grounds in the city where suitable grassy communities
have evolved and are left undisturbed.

Artemisia vulgaris   Mugwort
1 2 3 4 5 6 7 8 9 10 11 12 13 14
Native, very common.
Found in most waste grounds and many other disturbed
habitats throughout the city. It is not generally a pioneer
in colonising a waste ground, but persists once established.

A. absinthium   Wormwood
— — — — 5 — 7 8 — — — — — —
Introduced, rare.
The first record of this plant in County Dublin was in
1799 and it was probably introduced as a pot herb for

124

medicinal use. The plant still occurs in the region of Christ Church Cathedral where Colgan noted it in 1904. Colgan believed it to be a relic of window gardening in tenement houses now demolished. It does not seem to be spreading much except to a new site in zone 8. Church yard, Werburg Street, (5), '80, DNFC. Waste ground off Bridge Street, (7), '80, DNFC. Edge of canal, Sheriff Street, (8), '80, DNFC.

Tussilago farfara   Colt's Foot
1 2 3 4 5 6 7 8 9 10 11 12 13 14
Native, abundant.
Common in waste grounds, on roadsides and banks, usually preferring a habitat with a damp clay soil.

Petasites fragrans   Winter Heliotrope
– 2 – – – – 7 – – – 11 – 13 –
Introduced, occasional.
Frequently grown in gardens for its fragrant winter flowers and often escapes to become naturalised on shady banks and roadsides. It grows commonly on the banks of the Grand Canal (2). Grounds of St Brendan's Hospital, (11), '81, MSS, PWJ. Disused Cattle Market, Aughrim Street, (13), '81, DNFC.

Senecio jacobaea   Ragwort, Ragweed
1 2 3 4 5 6 7 8 9 10 11 12 13 14
Native, abundant.
Found growing abundantly in most city habitats, including roadsides, canal banks, waste grounds and occasionally on walls. A robust plant that can survive considerable competition in its habitat.

S. aquaticus   Marsh Ragweed
– 2 – – – – – – – – – – – –
Native, rare.
Found growing on the Grand Canal bank only, '81, DNFC.

125

S. squalidus
Oxford Ragwort

1 2 3 4 5 6 7 8 9 10 11 12 13 14
Introduced, very common.
Introduced in Ireland c. 1800. It was first recorded in Cork, then it appeared on the walls of the Liberties in Dublin, presumably introduced from Cork by rail traffic. Now it grows throughout Dublin on walls and in waste places and is often one of the commonest weeds to be seen. It was still rare in Dublin in 1961 (DNFC Supplement). (Figure p. 4)

S. viscosus
Sticky Groundsel

1 — — — — — — — — 10 - — — —
Introduced, very rare.
Found in Dublin during the last century, it disappeared up to the time of the present study, when during 1981 two specimens were found. It is quite an easy plant to recognise as it is sticky and generally covered in debris, dirt and seeds that have blown onto it. Growing in a waste ground near St Andrew's School, Pearse Street, (1), '81, PWJ, and in a waste ground off Dominick Street, near the Parnell Street end, (10), '81, PWJ.

S. vulgaris
Groundsel

1 2 3 4 5 6 7 8 9 10 11 12 13 14
Native, abundant.
Grows as a weed of waste grounds, roadsides, gardens, walls and the canal banks. It prefers a soil which has recently been disturbed or cultivated or which at least is not hard and compacted.
In zones (4) and (13) a rayed variant var. *hibernicus* of this species has been found. It may be a hybrid between this species and *S. squalidus*, which is recorded in a number of counties in Ireland. A complex hybrid (called a natural allopolyploid) derived from this is recorded in North Wales, and named *S. cambrensis*, but Dublin plants are unlikely to be this. Forms with ligulate flowerheads have been named subsp. *denticulatus*, but these also may be the result of introgressive hybridisation. So the Dub-

lin rayed groundsels may be a rare rayed form of the common *S. vulgaris* or the result of hybridisation between this species and *S. squalidus*. Outside Tara Street Baths, (4), '79, JRA. bases of walls on Hawkins Street, (4), '80, DNFC. Old Cattle Market, Aughrim Street, (13), '81, PWJ, JRA.

Calendula officinalis   Pot Marigold

$--3-------------$

Introduced, rare.

A commonly grown bedding annual of suburban gardens that frequently seeds itself nearby. It is not often planted in gardens in the city which may explain its scarcity.

Carlina vulgaris   Carline Thistle

$-------------14$

Native, very rare.

Its only known locality in the city is on a railway embankment near Cross Guns Bridge on the Royal Canal bank where it occurs with other calcareous grassland species. It is a calcicole of dry banks, sandhills and gravelly places.

Arctium minus   Burdock

1 2 3 4 − 6 7 8 9 10 11 12 13 14

Native, common.

A weed of waste grounds and particularly long-established ones. Also found on roadsides and on the canal banks. The plant has a good seed dispersal mechanism which helps it spread easily to new open ground. Its globular heads bear numerous bracts each with a hooked point.

Carduus tenuiflorus

$-------------14$

Native, very rare.

A plant of dry habitats. Growing in a car park near Cross Guns Bridge on the Royal Canal Bank, '81, DD.

Cirsium vulgare   Spear Thistle

1 2 3 4 5 6 7 8 9 10 11 12 13 14

Native, abundant.
Grows abundantly in many habitats in the city, in waste grounds, roadsides, derelict gardens and the canal banks. Its light seed is easily dispersed by wind.

C. palustre Marsh Thistle
1 2 — — — 6 — — — — — — — —
Native, rather rare.
Grows on the Grand Canal bank and in a few other marshy waste grounds. Apparently absent from the Royal Canal in the city.

C. arvense Creeping Thistle
1 2 3 4 5 6 7 8 9 10 11 12 13 14
Native, abundant.
Grows in dry waste ground and roadsides. It is a troublesome weed of gardens and can easily regenerate itself from a small piece of root. It is a very variable species especially with regard to leaf shape. A number of different subspecies have been described.

Centaurea nigra Knapweed, Black Heads
1 — — 4 — — — 8 — — 11 12 13 14
Native, locally frequent.
A plant of closed grassy communities. It grows in pastures, roadsides and grassy banks preferring a dry habitat.

Hypochoeris radicata Cat's Ear
— 2 3 4 — 6 7 8 — — 11 12 13 14
Native, frequent.
A plant of grassy communities, usually where the grass is short, or waste grounds or roadsides. It is absent from some of the more built-up zones of the city. Easily recognised by the short pointed 'cat's ear'-like bracts on the stems.

Leontodon autumnalis Hawkbit
— — — — — — — 8 — — — 12 — 14
Native, occasional.

Very like the previous species but is less coarse and hairy in appearance. It is a plant of damp grassland. Railway embankment near Cross Guns Bridge on the Royal Canal, (14), '81, DNFC. Lawn outside the Custom House on Custom House Quay, (8), '81, DD, JRA, PWJ.

Tragopogon pratensis Goat's Beard
$- - - 4 - - 7\ 8 - - - - - 14$
Native, occasional.
A weed of dry banks and some waste places. It is not common in the inner city. Waste ground off Moss Street, (4), '79, DNFC; near canal at Sheriff Street, (8), '80, DNFC. Railway embankment at Cross Guns Bridge, (14), '81, DNFC.

Sonchus asper Sow Thistle
$1\ 2\ 3\ 4 - 6 - - - 10\ 11\ 12\ 13\ 14$
Native, occasional.
A plant of waste grounds and roadsides. A difficult species to determine with certainty, it is variable and many specimens resemble the next species, *S. oleraceus*. Further work is needed on the species in Dublin.

S. oleraceus Sow Thistle
$1\ 2\ 3\ 4\ 5\ 6\ 7\ 8\ 9\ 10\ 11\ 12\ 13\ 14$
Native, abundant.
A very common plant in most ruderal habitats in the city, growing in waste grounds, roadsides, cracks in the pavements, gardens and even occasionally on walls. Like *S. asper*, it is very variable and some specimens approach this species in morphology.

S. arvensis Sow Thistle
$1 - - 4 - - 7 - - 10 - 12\ 13\ 14$
Native, occasional.
Found in waste grounds throughout the city. It is not common anywhere but occurs regularly. Its large yellow flowers and tall habit make it a distinctive plant.

129

Mycelis muralis   Wall Lettuce
– – 3 – – – – – – – – – – –
Introduced, very rare.
Colgan noted this plant as locally abundant in parks and
demesnes in Dublin in 1904. It may be spreading and
easily colonises walls and gravelly ground. Found in a
single locality in the inner city, beside the pavement on
Merrion Square, '80, PHC.

Taraxacum officinale agg.   Dandelion
1 2 3 4 5 6 7 8 9 10 11 12 13 14
Native, abundant.
This plant grows in most habitats in the city. The genus
Taraxacum is a very difficult group taxonomically and it
is almost impossible for anyone but an expert to determine species. Plants collected late in the season have
their distinctive characters masked and are impossible to
identify. A collection of *Taraxacum* specimens was made
in the city as a representative sample of those present.
Apart from these no attempt was made to determine
species.

Lapsana communis   Nipplewort
1 2 3 4 5 6 7 8 9 10 11 12 13 14
Native, very common.
Occurs in many waste grounds and ruderal habitats
throughout the city. It has spread from woods and
hedges onto cultivated and waste ground.

Crepis capillaris   Smooth Hawksbeard
– – 3 4 – – 7 8 – – 11 12 13 14
Native, frequent.
Grows in waste places, walls, roadsides and grassy habitats throughout the city. This species has achenes which
are not beaked.

C. vesicaria (*C. taraxacifolia*)   Beaked Hawksbeard
1 2 3 4 5 6 7 8 9 10 11 12 13 14
Introduced, very common.

130

Occurs on dry banks, in waste grounds, wall tops and pastures. Prefers a calcareous soil. Very common throughout the city. There are a number of European subspecies of this plant but the one that occurs in Ireland is subspecies *haenseleri*. This species differs from the last by having achenes which are beaked and the involucral bracts have white membranous edges.

Hieracium gougetianum   Hawkweed
— — — — — — — — — — — 12 — 14
Introduced, rare.
Occurs on the Royal Canal and on some railway embankments in the northern part of the city. This species is native to the Pyrenees. It is not known how it was introduced into Dublin. This genus is very variable and diverse because like *Rubus* and *Taraxacum* it is apomictic. There appears to be only a single species in the inner city.

MONOCOTYLEDONES

ALISMATACEAE

Sagittaria sagittifolia   Arrowhead
— 2 — — — — — — — — — — — —
Native, rare.
A plant of canals, rivers and lakes in Ireland. A few plants occur near Baggot Street Bridge on the Grand Canal. It seems to have declined in recent years in the city part of the Canal probably due to competition from *Ceratophyllum demersum*, '81, MS.

Alisma plantago-aquatica   Great Water Plantain
— 2 — — — — — — — — — — —
Native, rare.
Occurs only in the Grand Canal in Dublin. In Ireland it is a plant of ditches, ponds and other wet places, usually growing on a muddy substrate.

HYDROCHARITACEAE

Elodea canadensis   Water Thyme, Canadian Pondweed
— 2 — — — — — — — — — — — 14
Introduced, locally frequent.
Occurs commonly as a water plant on the Royal and
Grand Canals. This plant was first naturalised in Ireland
in 1834, then in England 25 years later, where it choked
the River Cam, at Cambridge. It is often a serious weed
and commonly clogs waterways in Ireland.

POTAMOGETONACEAE

Potamogeton obtusifolius   Pondweed
— 2 — — — — — — — — — — — —
Native, rare.
This species was not found during the present study, but
a specimen collected in 1972, from the Grand Canal near
Baggot Street Bridge, is in the National Herbarium (**DBN**),
collected by L. Farrell.

P. crispus   Curled Pondweed
— 2 — — — — — — — — — — -- 14
Native, locally frequent.
An aquatic plant that is locally frequent above Emmet
Bridge on the Grand Canal. It can be distinguished by
the leaves which have finely serrate margins. In the Royal
Canal, near Dakota Printing works (14), '74, DD.

P. filiformis   Slender-leaved Pondweed
— 2 — — — — — — — — — — — —
Native, rare.
Found in the Grand Canal, east of Baggot Street Bridge.
In Europe it occurs generally in places close to the sea
and often in brackish water. '81, MS.

Groenlandia densa   Opposite-leaved Pondweed
— 2 — — — — — — — — — — — —
Native, rare.

This species is the only protected Irish species that occurs within the inner city. In Ireland it occurs very rarely in streams, ditches and canals. It was found in certain sections of the Grand Canal in Dublin. '67, MS, '81, JS, DD.

ZANNICHELLIACEAE

Zannichellia palustris
— 2 — — — — — — — — — — — — —
Native, rare.
A plant that occurs in fresh or brackish water that is frequent near the coast in Ireland but rare inland. Found only in the Grand Canal in the city. Occurs locally above Portobello Bridge, '76, JS; near Baggot Street Bridge, '81, MS.

LILIACEAE

Hyacinthoides non-scriptus (*Endymion non-scriptus*)

Bluebell

— — — — — — — — — — 12 — —
Native, rare.
This native plant is found in a single locality in the inner city occurring on a disused railway line near the Cabra Road, '81, PR.

H. hispanicus   Spanish Bluebell
— — — — — — — — — — — 13 —
Introduced, rare.
A garden plant that occasionally becomes naturalised in waste grounds. Found in a single station in the inner city outside of the garden environment in a waste ground near Montpelier Hill, '80, DNFC.

Allium vineale   Crow Garlic
— 2 — — — — — — — — 12 — 14
Native, occasional.
Occurs in the inner city on banks and damp ground. The plants in Dublin belong to one variety, var. *compactum*,

where the flowers are completely replaced by bulbils in the umbel. The var. *vineale* seems to be absent from the city. On a disused railway line near Cabra Road, (12), '81, PWJ, PR. Banks of the Grand Canal east of Charlemont Bridge, (2), '81, MSS. Banks of the Royal Canal, (14), '80, DNFC.

AMARYLLIDACEAE

Narcissus sp.   Daffodil, Narcissus

– 2 – 4 – – – – – – – – – –

Introduced, rare.

A difficult group taxonomically with numerous hybrids and cultivars. Occasionally bulbs are discarded which may become established in the wild. In a heap of soil beside the street, Charlemont Street, (4), '81, PWJ. A few specimens found growing on the Grand Canal bank near Charlemont Bridge, (2), '81, PWJ.

IRIDACEAE

Iris pseudacorus   Yellow Flag

– 2 – – – – – – – – – – – 14

Native, occasional.

A plant of wet habitats. Found growing on the banks of the Royal and Grand canals.

Tritonia x crocosmiflora   Montbretia

– 2 3 4 5 6 7 8 9 10 11 – 13 14

Introduced, frequent.

A garden plant that frequently becomes established in waste grounds and other habitats throughout the city. It is a hybrid that originated in cultivation between two native south African species and is capable of producing some fertile seed. A very commonly naturalised plant in some parts of the west and southwest of Ireland. It is still frequently grown in gardens.

134

Juncus inflexus   Rush
  $- 2 \ 3 - - 6 \ 7 - 9 - - 12 - 14$
  Native, frequent.
  Occurs quite frequently in damp places on the canal banks
  and in waste grounds. It is the commonest species of
  Juncus encountered in the inner city. It is a calcicole and
  may be more successful than other species in drier
  habitats.

J. effusus   Common Rush
  $1 - - - - - - - - - - - - - -$
  Native, very rare.
  One tussock of this plant was found growing on the
  Grand Canal Dock, '79, PWJ.

J. conglomeratus
  $- - - - - - - - - - - - 13 -$
  Native, very rare.
  A plant of damp habitats more usually found in upland
  regions and with a narrower ecological amplitude than
  the last species occurring on even more acid soils. One
  record only, growing in a waste ground near Montpelier
  Hill, '80, DNFC.

J. bufonius agg.   Toad Rush
  $- - - - - - - - - - - 12 - -$
  Native, very rare.
  This taxon is made up of an aggregate of species in
  Ireland. They are difficult to separate and this record was
  not determined to species. They are common in Ireland
  on muddy, sandy or heathy soils. A single station for the
  plant occurs in the inner city, on a disused railway line
  near Cabra Road, '81, PR, PWJ.

J. articulatus   Jointed Rush
  $- - - - - - - - - - - 12 - -$
  Native, very rare.

A plant of wet fields, roadsides, marshes and lake-shores in Ireland, found in one locality in the inner city, with the previous species on a disused railway line near Cabra Road, '81, PR, PWJ.

Festuca pratensis   Fescue
— 2 — — — — — — — — — — — — —
Native, very rare.
A plant that often occurs in dry habitats in Ireland and is found in one place on the Grand Canal bank, west of Portobello Bridge, '81, DNFC.

F. rubra   Red Fescue
1 2 3 4 5 6 7 8 9 10 11 12 13 14
Native, common.
Common in gardens, waste grounds and at the bases of walls. It usually occurs as small dense tufts which persist for many years. Probably often introduced as lawn seed.

F. ovina   Sheep's Fescue
— — — 4 — — — — — — — — — —
Native, rare.
Found in dry habitats. It may have been overlooked in some stations as it is difficult to distinguish from *F. rubra* in closely cut grass. College Park, Trinity College, '80, MSS, PWJ.

Lolium perenne   Perennial Rye Grass
1 2 3 4 5 6 7 8 9 10 11 12 13 14
Native, abundant.
Occurs in most city habitats in abundance. It is by far the commonest grass to be found in the city and is generally the main constituent of grass seed for lawns and pastures. (Figure p. 139)

L. multiflorum   Italian Rye Grass
1 2 — — — — 7 — — — — — 14
Introduced, occasional.

A widely cultivated forage grass that is occasionally found in ruderal habitats in the city. It resembles the last species closely except that each lemma has a long terminal awn.

L. perenne x multiflorum
A hybrid between these two species was found on the Grand Canal bank near Grand Canal Street Bridge. It is intermediate between the two species and has a short terminal awn on each lemma. (2), '80, PWJ, MSS.

Vulpia bromoides
— — — — — — — — — — 11 12 — —
Native, very rare.
Reported by Colgan as very frequent in 1904, it colonises wall tops and dry sandy areas. It is a slender annual that is not successful in the inner city. On a dry gravelly bank beside the railway line near Cabra Road, (12), '81, PR. St Brendan's Hospital, wall top (11), '81, DD.

Desmazeria marina
1 — — — — — — — — — — — — — —
Native, very rare.
An annual grass rarely found far from the sea, growing on walls and quays. Only one locality for the species in the city, growing in a waste ground at Sir John Rogerson's Quay, with *D. rigida*, '81, PWJ, DD, JRA, EL.

D. rigida
1 — — 4 — 6 — — — — 11 12 13 14
Native, frequent.
This species, like the last, is annual but is not as closely restricted to coastal regions. It occurs in waste grounds, on walls and pavement cracks throughout the city. There seems to be no discernable pattern to its distribution but it is absent from some parts and quite common in others. A distinct variant of this species, var. *majus* has been found in four zones (1, 4, 6 and 11). This plant is taller than the type and is larger in all parts. Comparative culti-

1. Aethusa cynapium
2. Heracleum sphondylium
3. Lolium perenne
4. Elymus repens
5. Bromus sterilis
6. Hordeum murinum

139

vation experiments suggest that this variety is a genetic variant rather than an environmentally controlled variant. Where the two varieties grow in mixed populations a complete range of intermediates occurs. There appears to be no difference in ecological preference between the two varieties.

Poa annua   Annual Meadow Grass
1 2 3 4 5 6 7 8 9 10 11 12 13 14
Native, abundant.
Occurs in a wide variety of habitats from waste grounds to walls, gardens and recently disturbed habitats. The name *'annua'* is a misnomer in many cases as many of the specimens found were certainly not annual. It is the commonest species of grass found on walls and growing in pavement cracks where it is tolerant of trampling.

P. trivialis   Rough Meadow Grass
1 2 3 4 5 6 7 8 9 10 11 12 13 14
Native, common.
This species and the next species are rather similar in appearance but can be readily distinguished using two easily seen characteristics. In this species the ligule is more or less pointed and the rootstock is not creeping. It is a plant of grassy waste places and banks. Occasionally it is found as tussocks in waste grounds without a grassy plant community. In this way it is similar to the next species. This may be the beginning of secondary colonisation of waste ground leading to a closed grassland community.

P. pratensis   Meadow Grass
1 2 3 4 5 6 7 8 9 10 11 12 13 14
Native, very common.
Found in all zones in the city. It is however commoner than the last species being more abundant in waste grounds than the former. This species has a short and truncated ligule and the rootstock is creeping.

Dactylis glomerata   Cock's Foot Grass
1 2 3 4 5 6 7 8 9 10 11 12 13 14
Native, very common.
Found in waste grounds, on roadsides and banks and
occasionally on walls. It is one of the more robust grass
colonisers of waste grounds and can develop large tussocks if left undisturbed and there is sufficient soil.

Cynosurus cristatus   Thraneen, Crested Dog's-Tail
1 — — 4 — — 7 — — — — 13 14
Native, rather rare.
Occurs very occasionally in waste places and on roadsides. It is an indicator of poor soils whereas many of the
city's derelict sites are nutrient-enriched from rubbish
dumping.

Briza media   Quaking-grass
— — — — — — — — — — — — — 14
Native, very rare.
Occurs in only one station in the inner city. Although it
is not strictly calcicole in general it is commoner on limestone soils and favours a dry habitat. On a railway embankment near Cross Guns Bridge and on the Royal Canal, as
part of a limestone grassland community.

Glyceria maxima
1 2 — — — — — — — — — 13 14
Native, locally common.
Almost entirely restricted to the canals where it is a common plant. There is a single record of this species growing
on the Liffey wall at Arran Quay, (13), '81, PWJ. Colgan
notes its abundance along the canals, the Liffey, Tolka
and parts of other south Dublin rivers.

Bromus sterilis   Barren Brome
1 — — — 5 6 7 8 9 — 11 12 13 14
Native, frequent.
Locally frequent in waste grounds and roadsides and
occasionally as a garden weed. It is a plant of dry habi-

141

tats. Colgan in 1904 found it was abundant mainly along the coast of County Dublin. (Figure p. 139)

B. ramosus
$-- 3 ------------$
Native, very rare.
A plant of woods and shady places and a rather un-expected find in the city. There is only a single station for the plant in the city, in a waste ground off Earlsfort Terrace, '81, PWJ, JRA. The site was an old school which was demolished quite recently. It was surrounded by a mature garden with trees and the plant may be a sur-vival from that garden.

B. hordeaceus (*B. mollis.*)   Soft Brome
$- 2 - 4 --- 8 ------$
Native, rather rare.
A plant of waste places, roadsides and meadows in Ireland. Found in the inner city in a few waste grounds and on the banks of the Grand Canal, (2), '80, MSS, PWJ.

Brachypodium sylvaticum
$-- 3 ------------$
Native, very rare.
A plant of woods and shady places that occurs in a waste ground off Earlsfort Terrace. This species is occasionally planted in gardens as ground cover in shade, which may well be its origin in this station. There are the remains of a garden nearby with large trees.

Elymus repens (*Agropyron repens*)   Scutch, Couch-grass
1 2 3 4 − 6 7 8 9 10 11 12 − 14
Native, common.
A common weed of waste places, spreading by means of underground rhizomes. It regenerates easily from frag-ments of rhizome often discarded with garden waste. It flowers rather rarely in Dublin and relies on its rapid vegetative spread for colonisation. (Figure p. 139)

Triticum aestivum   Wheat

1 2 — — — — — — 9 — — — 13 14

Introduced, frequent.

Plants of this common agricultural crop are found frequently throughout the city, occurring only as a casual weed. All records are probably derived from seed spilt from passing trucks carrying grain, as it is more commonly found on roadsides than in waste grounds. It does not appear to persist in the same places from year to year without fresh input of seed.

Hordeum distichon   2-Rowed Barley

1 2 — — — — 7 — 9 — — — — —

Introduced, occasional.

Cultivated as a cereal and occasionally plants are found in the inner city arising from spilled seed, usually on roadsides.

H. vulgare   6-Rowed Barley

— 2 — — — 6 — — — — — — — —

Introduced, occasional.

Also a cultivated cereal like the last species, but slightly less common. It also arises from spilled seeds and occurs on roadsides and occasionally in waste grounds.

H. murinum   Wall Barley

1 2 3 4 5 6 7 8 9 10 11 12 13 14

Introduced, abundant.

Although introduced, this annual is one of the commonest weeds of the city. It is found growing in waste grounds, on roadsides, on walls, in the cracks of pavements and most other ruderal city habitats. Oddly it is rare in many other parts of Ireland away from Dublin, where it is a very characteristic plant. It seeds itself freely in the city and often forms pure swards in waste grounds. It is native to much of Europe, including Britain. (Figure p. 139)

Avena fatua   Wild Oat

1 — — — — — 7 — 9 — — — — —

Introduced, occasional.

A weed of cultivated crops, usually with crop plants with which it probably was harvested. Similar to the next species, it differs in that it has a tuft of orange hairs at the base of the lemma and the seeds are shed readily from the spikelets. In the next species, *A. sativa*, the tuft of lemma hairs is absent and the seeds are retained in the spikelets.

A. sativa   Cultivated Oat

1 — — — — — 8 9 — — — — —

Introduced, occasional.

Cultivated as a cereal crop. Grows in Dublin in waste grounds and on roadsides where it has fallen from passing trucks carrying grain. It does not seem to persist from year to year in the same place and does not become naturalised.

Avenula pubescens (*Helictotrichon pubescens*)

— — — — — — — — — — — — — 14

Native, very rare.

A calcicole plant of dry limestone soils or gravel banks. Occurs with other limestone grassland species on one railway embankment near Cross Guns Bridge on the Royal Canal, '81, DNFC.

Arrhenatherum elatius   False Oat

1 2 3 4 5 6 7 8 9 10 11 12 13 14

Native, very common.

Very common in waste grounds, on the tops of walls and in old derelict gardens where it tends to become the dominant grass species with *Dactylis glomerata*. It seeds itself freely in all habitats in which it occurs.

Trisetum flavescens

— — — — — — — — — — — — — 14

Native, very rare.

Railway embankment west of Cross Guns Bridge, on the Royal Canal, '82, DNFC.

Lagurus ovatus

— — — 4 — — — — — — — — — — —

Introduced, very rare.

Sometimes grown as an ornamental grass. One plant in Kildare Street, '80, MS.

Anthoxanthum odoratum   Sweet Vernal Grass

— — — — 5 — — — — — — — — 14

Native, rare.

In Ireland this is a plant of pastures, heaths, meadows and woods, but it is rare in the inner city of Dublin. It has been found only on the banks of the Royal Canal, (14), '80, DNFC, and in a waste ground near the South Circular Road, (5), '80, DNFC.

Holcus lanatus   Yorkshire Fog

1 2 3 4 5 6 7 8 9 10 11 12 13 14

Native, very common.

Grows in most habitats in the city, in waste grounds, on walls and roadsides and as a garden weed. It occasionally occurs in the cracks of pavements. One of the commonest grasses of the city.

H. mollis

— — 3 4 — 6 — — — — — — — —

Native, rather rare.

Rather similar to the last species and may have been overlooked in one or two localities, but is nowhere common. It differs from *H. lanatus* in that the stems arise singly from a creeping rhizome and the nodes are much hairier than the internodes. It does not form dense tussocks like the previous species. Found in waste grounds and once as a flower-bed weed. Edge of Botany School, Trinity College, in a flower-bed, (4), '80, MSS, PWJ. Waste ground off Clanbrassil Street Lower, (6), '81, JP. Waste ground off Mount Street Lower, '80, KD, PWJ.

Agrostis capillaris (*A. tenuis*)   Bent Grass

1 — 3 — — — — — 9 — 11 — 13 —

Native, occasional.
In Ireland this plant is one that grows in dry pastures, heaths, on mountain tracks and at the bases of walls throughout the city. It is rarely found in waste grounds or on open ground. Many of these plants are rather enigmatic in that they correspond closest but not entirely with descriptions of this species. Further work is desirable to establish their status. The city habitat is very different from their usual environment of acid grasslands.

A. stolonifera   Bent Grass
1 2 3 4 5 6 7 8 9 10 11 12 13 14
Native, abundant.
A very common plant throughout the city occurring in waste grounds and on roadsides. It can easily be recognised by the presence of numerous leafy stolons spreading from the base of the stems that on most occasions clothe the ground. It prefers a damp habitat.

Phleum pratense   Timothy-grass, Cat's Tail
1 2 — — — — 7 8 — — 11 — 13 14
Native, frequent.
In Ireland it grows in pastures, meadows and on roadsides. In Dublin it is frequent only in long standing grassy communities which occur in a number of places including the canal banks. It is not a plant that is often seen in the inner city waste grounds.

Alopecurus pratensis   Meadow Foxtail
— 2 — — — — — — — — — — 13 14
Native, occasional.
The preferred habitat of this plant is damp meadows and pastures. In the city it occurs on the canal banks and in a few other damp grassy communities.

Phalaris arundinacea   Reed Grass
1 2 — — — 6 — — — — — — — 14
Native, occasional.
A plant of ditches and streamsides that in Dublin occurs

in the Royal Canal (14) and the Grand Canal (1 and 2) and in one other locality on a roadside, near Dolphin's Barn Road. This latter station is not a damp habitat so it was an unusual place for the plant to occur, (6), '81, DNFC.

P. canariensis   Canary-grass
1 − − 4 − 6 7 − − 10 11 − − −

Introduced, occasional.

This plant arises from seed which is the main constituent of canary bird-seed. All of the records are certain to have arisen from discarded seed that has grown in the wild. It is quite a common plant of rubbish dumps for this reason. In the city it was found in waste grounds and growing in the cracks of pavements. It is purely a casual and probably never persists in one place for more than a year.

LEMNACEAE

Lemna trisulca   Ivy-leaved Duckweed
1 2 − − − − − − − − − 12 − 14

Native, occasional.

Found in four localities in the city. In ponds in a disused railway line near Cabra Road, (12), '81, PR, PWJ. Beside the Grand Canal Street Bridge, with the following species, (1), '81, JRA, PWJ. Grand Canal near Wilton Place, (2), '81, MS. Common on the Royal Canal, (14), wherever the water is still. Most of the Grand Canal has water which is too fast-moving for this species.

L. minor   Common Duckweed
1 2 − − − − − − − − − 12 − 14

Native, occasional.

Most of the Grand Canal has water which is too fast-flowing for this species. It is common on the Royal Canal, (14), wherever the water is still. Found in ponds in a disused railway line near Cabra Road, (12), '81, PR, PWJ. Beside the Grand Canal Street Bridge, with the previous species, (1), '81, JRA, PWJ.

147

Spirodela polyrhiza

$- - - 4 - - - - - - - - 12 - -$

Native, rare.

This species is similar to the last, but is larger and has a tassel of several roots from each frond rather than one as in the case of *L. minor*. Like the last species it will not survive in flowing water. In a water tank, behind the Zoology Building, Trinity College, (4), '80, MSS. In ponds in a disused railway line near Cabra Road, (12), '81, PR, PWJ.

SPARGANIACEAE

Sparganium emersum   Bur-reed

$1\ 2 - - - - - - - - - - - 14$

Native, locally frequent.

This aquatic plant grows in a number of localities in both the Grand and Royal Canals. It is a plant of slow streams, ditches and lake-margins.

TYPHACEAE

Typha latifolia   Bulrush

$1 - - - - - - - - - - - - -$

Native, very rare.

Although this plant is aquatic and grows in ditches, slow streams, marshes and lake-margins in Ireland, it is absent from both canals in the inner city. The single station for the plant is a bizarre one, in a water tank, in Pearse Station on Westland Row, '79, DNFC.

CYPERACEAE

Carex otrubae   Sedge

$- - - - - - - - - - - - - 14$

Native, very rare.

It is an unexpected plant but is found in coastal localities all around Ireland, especially by salt marshes and prefers brackish conditions. It is very rare inland. One station for this plant in the city, growing with *Aster tripolium* near the mouth of the Royal Canal, close to Sheriff Street, on mud. '81, JRA, PWJ, DD, EL.

C. ovalis

$- - - - - - - - - - - - 13 -$

Native, very rare.

This, along with other species in this waste ground, is probably a relic of rough pasture and hedgerows. In a large waste ground near Montpelier Hill, '80, DNFC. Growing in a damp habitat with *Carex hirta.* A study of old maps of Dublin indicated that area has never been greatly disturbed or built upon.

C. hirta

$- 2 - - - - - - - - - - 13 -$

Native, rare.

Frequent in damp pastures and marshes. In the inner city it was found on the banks of the Grand Canal, near Charlemont Bridge, (2), '81, MSS, and in a large waste ground off Montpelier Hill, (13), '80, DNFC.

C. flacca

$- - 3 - - - - - - - - - - -$

Native, rare.

This is a common sedge of calcareous grasslands, marshes and damp woods, but appears not often to colonise waste places or disturbed ground. Found in a single locality in the inner city in a waste ground off Montpelier Hill, '79, DH, PWJ.

ORCHIDACEAE

Gymnadenia conopsea   Fragrant Orchid

$- - - - - - - - - - - - - 14$

Native, very rare.

Railway embankment west of Cross Guns Bridge on the Royal Canal, '82, DNFC.

Dactylorhiza fuchsii   Common Spotted Orchid
— — — — — — — — — — — — 14
Native, very rare.
Although this is the commonest species of orchid in
Ireland, it was a pleasant surprise to find it in the city
with the next species. Orchids are an unexpected feature
of the urban environment. It is commonest on dry calcareous soils including railway embankments. Railway
embankment near Cross Guns Bridge, by the Royal
Canal, '80, DNFC.

Anacamptis pyramidalis   Pyramidal Orchid
— — — — — — — — — — — — 14
Native, very rare.
Grows in a similar habitat to the last species. It is a calcicole plant of dry banks, pastures, gravelly places and
sandhills. Found with the last species on a railway embankment near Cross Guns Bridge, by the Royal Canal, '80,
DNFC.

BUDDLEJA DAVIDII

151

Check-list

The following is a check-list of the commoner plants of the city given in the form of a list of abbreviated scientific names. The majority of species which may be encountered in the city are included. When photocopied, the list will make a convenient card for botanists to use when exploring the waste grounds and other habitats of the city.

Acer
Achill mill
Aesculus
Aethusa
Agrost cap
    stol
Allium vin
Alnus glut
Alopec prat
Anagal arv
Angelica
Anthoxanthum
Anthriscus
Antirrhinum
Apium nod
Arabidopsis
Arctium min
Arenaria
Arrhenath
Artemis abs
    vul
Asplen trich
    ru-mu
Aster trip
Atriplex has
    pat
Avena fat
    sat
Barbarea vul
Bellis

Brassica nig
    oler
    rapa
Bromus hord
    ster
Buddleja
Calysteg sep
    syl
Capsella
Cardam hirs
Carex hirta
Cent nigra
Centranthus
Cerast font
    glom
Ceratophyllum
Chamomilla
Cheiranthus
Chenop alb
Circaea lut
Cirsium arv
    vul
Convolvulus
Coron did
    squa
Corylus
Crataegus
Crepis cap
    ves
Cymbalaria

Cynosurus
Dactylis
Daucus
Desmaz mar
    rig
Diplotaxis
Digitalis
Dryopt fil-m
Elodea
Elymus rep
Epilob aden
    ang
    hirs
    mont
    obsc
    pal
    parv
Equiset arv
Eupatorium
Euphorb hel
    pep
Fallopia
Festuca rub
Filipendula
Fraxinus
Fumaria bast
    mur
    off
Galium apar
Geran diss

    rob
Geum urb
Glycer max
Hedera
Heracl sph
Hieracium goug
Holcus lan
    mol
Hordeum dist
    mur
    vul
Hyperic andr
    caly
Hypochoeris
Ilex
Impatiens gla
Iris pseud
Juncus inflex
Laburnum anag
Lamium pur
Lapsana
Lemna min
    tris
Leont aut
Leucanth vul
Ligustrum
Linaria pur
Lolium mult
    per
Lotus corn

Lycopersicon
Lycopus eur
Malva syl
Matric mar
    perf
Medicago lup
Melil altis
Mentha aquat
    suav
Mercur ann
Myosot arv
Nastur mic
    off
Nuphar
Odontites
Papaver dub
    lec
    rhoe
    somn
Parietaria
Partheno ins
    quin
Petasites fra
Phalaris arun
    can
Phleum prat

Phyllitis
Plantago lanc
    maj
Poa annua
    prat
    triv
Polyg amph
    aren
    avic
    lapat
    pers
Polypod vul ag
Potent ans
    rept
Prunella
Pteridium
Ranunc acr
    bulb
    fic
    rep
    scel
Raphan raph
Rapist rug
Reseda luteo
Reynout jap
Rosa can

Rubus frut ag
    idaeus
    ulmif
Rumex acella
    congl
    crisp
    obtus
    sang
Sagina apet
    proc
Salix atrocin
    cap
Sambucus nig
Scroph nod
Sedum acre
Senecio jaco
    squa
    vul
Silene vul
Sinapis arv
Sisymb off
    orie
Smyrnium
Solanum dulc
    tuber
Soleirolia

Sonch arv
    asp
    oler
Spargan emer
Spergul arv
Stellar med
Tanacetum par
    vul
Tarax off ag
Tragopogon
Trifol dub
    prat
    rep
Triticum aest
Tritonia
Tussilago
Ulmus glab
    proc
Urtica dioica
    urens
Verbascum
Veronic fili
    pers
Vicia sat
    sep

Bibliography

Contains those papers and books referred to in the text as well as others of relevance, together with a number of useful general texts (marked *).

Akeroyd, J.R. 1980. Two alien legumes from waste ground in Dublin. *Bulletin of the Irish Biogeographical Society* 4, 50.

Akeroyd, J.R. 1982. *Senecio viscosus* L. in Ireland. *Irish Naturalists' Journal* 20, 361-364.

*Allen, D.E. 1978. *The Naturalist in Britain — A Social History*. Penguin, Harmondsworth.

Bailey, K.S. (Afterwards Lady Kane). 1833. *The Irish Flora: Comprising the phaenogamous plants and ferns*. Hodges and Smith, Dublin.

Barkman, J.J. 1969. The influence of air pollution on bryophytes and lichens. In *Air Pollution*. Proceedings of 1st European Congress on the influence of air pollution on plants and animals, 197-209, Wageningen.

Brunker, J.P. 1944. *Flora of St James's Gate*. Ms. in the possession of H.J. Hudson.

Brunker, J.P. 1950. *Flora of the County Wicklow*. Dundalgan Press, Dundalk.

Campbell, J.L. The Tour of Edward Lhwyd in Ireland in 1699 and 1700. *Celtica* 5, 218-228.

Carvill, P.H. and Wyse Jackson, P.S. 1983. *Galinsoga ciliata* (Raf.) Blake in Ireland. *Irish Naturalists' Journal* 21, 31.

Chater, A.O. 1974. The street flora of central Aberystwyth. *Welsh Regional Bulletin* 21, 2-17. *Botanical Society of the British Isles*.

*Clapham, A.R., Tutin, T.G. and Warburg, E.F. 1962. *Flora of the British Isles*. 2nd Ed. Cambridge University Press, Cambridge.

Clinch, P. 1981. Botany and the Botanic Gardens. In *R.D.S., The Royal Dublin Society 1731-1981*, ed. J. Meenan and D. Clarke. Gill and Macmillan, Dublin.

Colgan, N. 1904. *Flora of the County Dublin.* Hodges Figgis & Co., Dublin.

Colgan, M. and Scully, R.W. 1898. *Cybele Hibernica* 2nd ed., Ponsonby, Dublin.

Collins, J. 1978. *Life in Old Dublin.* Cork.

Cosgrave, D. 1977. *North Dublin.* (First published in 1909) Dublin.

Desmond, R. 1977. *Dictionary of British and Irish Botanists and Horticulturalists.* Taylor and Francis, London.

Dublin Naturalists' Field Club. 1961. *A Supplement to Colgan's Flora of the County Dublin.* Government Publications, Dublin.

Ewen, A.H. and Prime, C.T. 1975. *Ray's Flora of Cambridgeshire.* Weldon and Wesley, Hitchin.

Fenton, A.F. 1964. Atmospheric pollution of Belfast and its relationship to the lichen flora. *Irish Naturalists' Journal* 14, 237-245.

*Fitter, R., Fitter, A. and Blamey, M. 1974. *The Wild Flowers of Britain and Northern Europe.* Collins, London.

Fraser, A.M. 1960. The Molyneux Family. *Dublin Historical Record* 16, 9-15.

Gibson, T.B. 1896. The Botany of a school playground in the heart of Dublin. *Irish Naturalist* 5, 277-284.

Gilbert, O.L. 1968. Bryophytes as indicators of air pollution in the Tyne Valley. *New Phytologist* 67, 15-30.

Grigson, G. 1975. *The Englishman's Flora.* Paladin, St. Albans.

Hawksworth, D.L., James, P.W. and Coppins, B.J. 1980. Checklist of British lichen-forming, lichenicolous and allied fungi. *Lichenologist* 12, 1-115.

Hill, M.O. 1979. New vice-county records and amendments to the Census Catalogue. *Bulletin of the Bryological Society* 34, 27.

Hoppen, K.T. 1970. *The Common Scientist in the Seventeenth Century.* Routledge and Keegan Paul, London.

Hughes, J.L.J. 1961. A tour through Dublin city in 1782. *Dublin Historical Record* 17, 2-12.

Joyce, W. St. J. 1977. *The Neighbourhood of Dublin.* (First published 1912) Gill and Son, Dublin.

*Keble Martin, W. 1965. *The Concise British Flora in Colour.* Ebury Press and Michael Joseph, London.

McCall, P.J. 1894. *In the Shadow of St. Patrick's.* A paper read before the Irish National Literary Society, April 27, 1893. (Reprinted 1976, Dublin)

*McClintock, D. and Fitter, R.S.R. 1956. *Collins' Pocket Guide to Wild Flowers.* Collins, London.

MacGiolla Phadraig, B. 1948. *Speed's Plan of Dublin.* Pt. 1. *Dublin Historical Record* 10, 89-96.

MacGiolla Phadraig, B. 1949. *Speed's Plan of Dublin.* Pt. 2. *Dublin Historical Record* 10, 99-105.

Mackay, J. 1806. A Systematic Catalogue of Rare Plants found in Ireland. *Transactions of the Dublin Society* 5, 121 *et seq.*

Mackay, J.T. 1825. Catalogue of the indigenous plants of Ireland. *Transactions of the Royal Irish Academy* 14, 103-198.

Mackay, J.T. 1836. *Flora Hibernica.* Curry and Co., Dublin.

MacLoughlin, A. 1979. *Guide to Historic Dublin.* Gill and Macmillan, Dublin.

MacLysaght, E. 1979. *Irish Life in the Seventeenth Century.* (First published in 1939) Irish Academic Press, Dublin.

Mitchell, M.E. 1974. The Sources of Threlkeld's Synopsis Stirpium Hibernicarum. *Proceedings of the Royal Irish Academy* 74B, 1-6.

Mitchell, M.E. 1975. Irish Botany in the seventeenth century. *Proceedings of the Royal Irish Academy* 75B, 275-284.

Molyneux, T. 1726. Appendix to Threlkeld's Synopsis Stirpium Hibernicarum. Davys, Norris and Worrall, Dublin.

Moore, C.C. 1976. Factors affecting the distribution of saxicolous lichens within a four kilometre distance of Dublin city centre. *Proceedings of the Royal Irish Academy* 76, 363-383.

Moore, D. and More, A.G. 1866. *Contributions towards a Cybele Hibernica, being the outlines of the geographical distributions of plants in Ireland.* Hodges, Smith and Co. Dublin.

More, A.G. 1872. On recent additions to the flora of Ireland. *Proceedings of the Royal Irish Academy* **2B**, 256-293.

Nelson, E.C. 1978. The publication date of the first Irish Flora: Caleb Threlkeld's Synopsis Stirpium Hibernicarum, 1726. *Glasra* **2**, 37-42.

Nelson, E.C. 1979a. "In the Contemplation of Vegetables" Caleb Threlkeld (1676-1728), his life, background and contribution to Irish botany. *Journal of the Society for the Bibliography of Natural History* **9**, 257-273.

Nelson, E.C. 1979b. Records of the Irish Flora published before 1726. *Bulletin of the Irish Biogeographical Society* **3**, 51-74.

Ní Lamhna, E., Richardson, D.H.S., Dowding, P. and Wells, J.M. 1983. *An Air Quality Survey of Cork City and Great Island carried out by Schoolchildren.* An Foras Forbatha, Dublin.

O'Dwyer, F. 1981. *Lost Dublin.* Gill and Macmillan, Dublin.

Ordnance Survey. 1917. Dublin Sheet 18. Geological Survey of Ireland.

Praeger, R. Ll. 1901. Irish Topographical Botany. *Proceedings of the Royal Irish Academy* (series 3), **7**, i-clxxxviii, 1-410.

Praeger, R. Ll. 1920. Ferns in Dublin City. *Irish Naturalist* **29**, 108.

Praeger, R. Ll. 1937. *The Way that I Went.* Hodges Figgis and Co., Dublin.

Praeger, R. Ll. 1949. *Some Irish Naturalists.* Dundalgan. Dundalk.

Pulteney, R. 1777. Memoir relating to Dr. Threlkeld. *Gentleman's Magazine.* **47**, 63-64.

Rocque, J. 1757. *Map of Dublin and Environs.*

Richardson, D.H.S. 1981. *The Biology of Mosses.* Blackwell, London.

*Salisbury, E. 1961. *Weeds and Aliens.* New Naturalist Series. Collins, London.

Scannell, M.J.P. 1971. *Ceratophyllum demersum* L. in County Dublin. *Irish Naturalists' Journal* **17**, 61.

Scannell, M.J.P. 1979a. A 17th century *Hortus Siccus* made in Leyden the property of Thomas Molyneux, at **DBN**. *Irish Naturalists' Journal* **19**, 320-21.

Scannell, M.J.P. 1979b. *Epilobium angustifolium* L. in the flora of Dublin. *Irish Naturalists' Journal* **19**, 327.

Scannell, M.J.P. and Synnott, D.M. 1972. *Census Catalogue of the Flora of Ireland*. Stationery Office, Dublin.

Shackleton, J.J. 1976. The effect of weed control on the vegetation of the Grand Canal with particular reference to *Ceratophyllum demersum* L. Unpublished Moderatorship Thesis, School of Botany, Trinity College, Dublin.

Smith, A.J.E. 1978. *The Moss Flora of Britain and Ireland*. Cambridge University Press, Cambridge.

Stephenson, P.J. 1948. Sean McDermott Street. *Dublin Historical Record* **10**, 83-88.

Synnott, D. 1981. *Epilobium adenocaulon* Hausskn. in Ireland. *Irish Naturalists' Journal* **20**, 234.

Synnott, D. 1982. An outline of the byrophytes of Meath and Westmeath. *Glasra* **6**, 1-71.

Threlkeld, C. 1726. *Synopsis Stirpium Hibernicarum*. Dublin.

Tutin, T.G., Heywood, V.H., Burges, N.A., Moore, D.M., Valentine, D.H., Walters, S.M. and Webb, D.A. edit., 1964-1980. *Flora Europaea* vols 1-5. Cambridge University Press, Cambridge.

Wade, W. 1794. *Catalogue Systematicus Plantarum Indigenarum in Comitatu Dublinense Inventarum*. Dublin.

Wade, W. 1804. Plantae Rariores in Hibernia Inventae. *Transactions of the Royal Dublin Society* **4**, i-xiv, 1-214.

Walsh, L. 1978. *Richard Heaton of Ballyskenagh*. Roscrea.

Warburton, J., Whitelaw, J. and Walsh, R. 1818. *History of the City of Dublin*. 2 vols. Cadell and Davies, London.

Webb, D.A. 1972. Two recent arrivals in County Dublin. *Irish Naturalists' Journal* **17**, 245.

*Webb, D.A. 1977. *An Irish Flora*. 6th Ed. Dundalgan Press, Dundalk.

Wright, G.N. 1825. *An Historical Guide to Ancient and Modern Dublin*. 2nd Ed. Baldwin, Cradock and Joy, London. (Reprinted Dublin 1980)

Wyse Jackson, P.S. 1982. Two records for *Senecio viscosus* L. in Dublin (H21). *Irish Naturalists' Journal* **20**, 507.

Wyse Jackson, P.S. 1982. Studying the flora of Dublin City. *Environmental Education Newsletter* **19**, 8-11. An Foras

Forbartha.

Wyse Jackson, P.S. 1984. *Rapistrum rugosum* L. in Ireland. *Bulletin of the Irish Biogeographical Society* 5, 15-18.

Glossary

Achene	A small, dry, single seeded fruit, e.g. dandelion seed.
Agrestal	A weed of cultivation.
Alien	A plant that is not native to Ireland.
Annual	A plant living for only one year.
Apomictic	-plant (Apomixis). Production of seeds without sexual fusion.
Awn	A bristle-like point borne on the bract(s) (q.v.) enclosing a grass flower.
Axil	The angle of the leaf and the stem.
Biennial	A plant which lives for two years.
Brackish	-water. Water which is slightly saline.
Bract	A small leaf or scale below the flower or inflorescence.
Bulbil	A small bulb or tuber arising in the axil of a leaf or in the inflorescence.
Calcareous	Containing calcium carbonate or limestone.
Calcicole	A plant preferring the presence of lime in the soil.
Calcifuge	A plant unable to tolerate the presence of lime in the soil.
Carpel	Single female unit of flower, consisting of ovary, style and stigma.
Casual	A plant occurring by chance, not regularly or permanently.
Composite	-flower. A group of flowers clustered into one head so that together they resemble one flower, e.g. dandelion or daisy.
Corolla	A collective term for the petals.
Critical	-group. A group of plants that is difficult to classify into species or to name, because

differentiating characters are obscure or variation is large between individuals due for example to apomixis (q.v.).

Cultivar — A cultivated variety of a species.

Drupe — A fleshy fruit containing a stone, e.g. cherry or plum.

Esker — A dry gravel ridge, deposited during the ice age by rivers running under the ice mass.

Excurrent nerve — A nerve, or central strand which extends beyond the blade of a leaf.

Exotic — -plant. A non-native and usually garden plant.

Floret — A single flower in an inflorescence of many flowers massed together, e.g. in Compositae.

Frond — The leaf of a fern.

Genus — A group of species with common characteristics and name, e.g. *Polygonum*, the knotgrass genus (pl. genera).

Glabrous — Without hairs.

Glandular hair — A hair with a sticky globular head.

Glaucous — Of dull greyish green or blue colour.

Glume — One of a pair of chaff-like bracts surrounding a grass spikelet.

Hyaline — -point. A colourless or transparent point, often hair-like in appearance.

Hybrid — A cross between two different species of plant or animal.

Inflorescence — A flowering branch, a more or less compact group of flowers, considered collectively.

Introgressive hybridisation — Where a range of variation between two species is created by hybrids crossing back with their parent species.

Lemma — The outer of two bracts enclosing the flower of a grass. These are the chaff of cereals. The lemma sometimes bears an awn (q.v.).

Ligulate — Strap-shaped. Often used to describe one type of floret in members of the Compositae.

161

Ligule	A small membranous outgrowth occurring in the leaf axils of grasses.
Loculus	One valve of a seed capsule (as in the Cruciferae).
Naturalised	-plant. An alien which has established and propagates itself in the wild, thus appearing to be native.
Perennial	A plant which lives for many years.
Ruderal	A plant of waste places.
Rhizome	An underground or creeping stem resembling a root.
Rhizoid	Fine root-like fibres in mosses and liverworts, functioning as anchors for the plant and as simple water absorption structures.
Salt marsh	A marsh influenced by salt water. In Ireland these are coastal and regularly covered by the tide.
Serrate	Sharply toothed, like the edge of a saw.
Siliqua	An elongated seed capsule (of Crucifers) which splits into two valves.
Species	The smallest generally-used category of classification of a plant or animal. Members of the same species have many characteristics in common and can interbreed. Members of different species generally cannot interbreed successfully.
Spikelet	A small, compact unit in a grass or sedge inflorescence, consisting of one or more flowers.
Stigma	The uppermost part of a carpel on which the pollen germinates.
Stolon	Creeping leafy shoot produced from the base of a plant.
Subspecies	A division of plants or animals below the species level.
Sucker	New shoot produced from the root of a (usually woody) plant, often some distance from the parent.
Taxonomy	The classification of plants and animals

	into genera, species, etc.
Tendril	A thread-like modified shoot or leaf by which some climbing plants cling.
Umbel	An inflorescence in which the flower stalks arise from one point, like umbrella-ribs, e.g. cow parsley or cowslip.
Variety	The class below subspecies of a plant or animal, involving a group which differs in one or more characteristics from the typical form and retains these in succeeding generations.
Vegetative reproduction	Asexual reproduction, thus not involving the production of seeds.

Index of Latin Names

167

Index of English Names

170

171

173

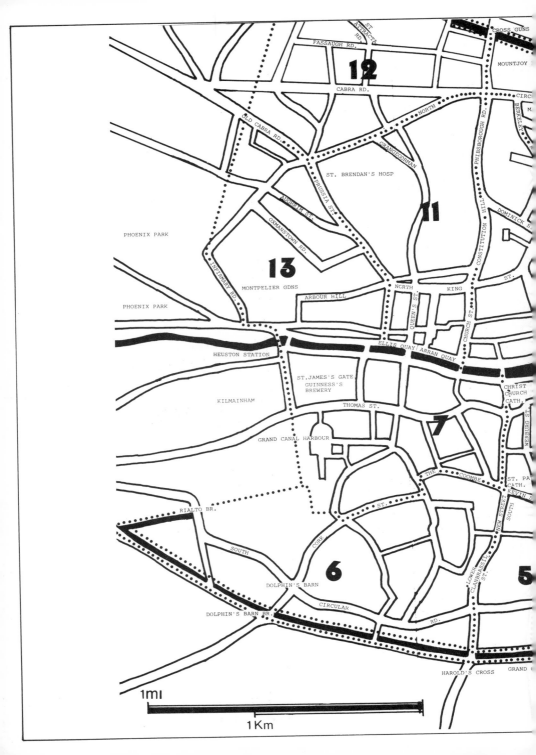

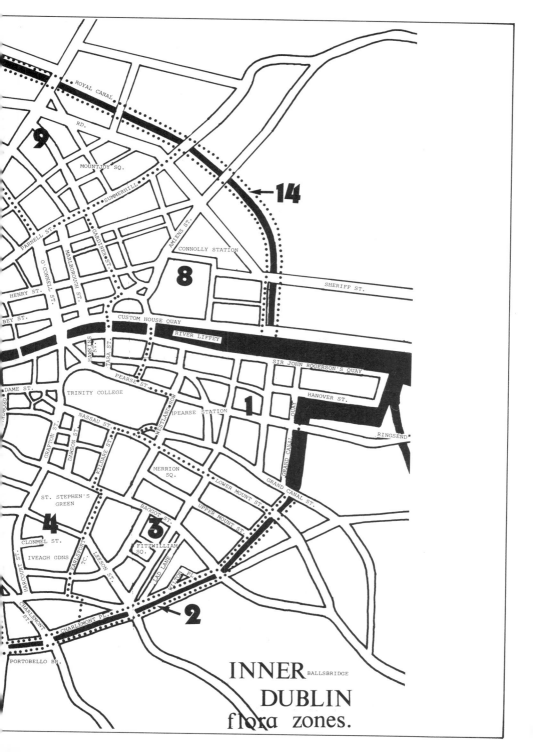

INNER DUBLIN
flora zones.